BECOMING DINOSAURS

A Prehistoric Perspective on Climate Change Today

WHAT'S HAPPENED BEFORE,
WHAT'S HAPPENING NOW,
AND WHAT WE MIGHT DO TO AVOID EXTINCTION

David Trexler

ISBN 10: 1-59152-092-4
ISBN 13: 978-1-59152-090-0

Published by TIMELINE DESIGNS, LLC

© 2012 by David Trexler

Cover and book design by Michael Berglund.

You may order extra copies of this book by calling Farcountry Press toll free at (800) 821-3874.

sweetgrassbooks
a division of Farcountry Press

Produced by Sweetgrass Books
PO Box 5630, Helena, MT 59604;
(800) 821-3874; www.sweetgrassbooks.com.

The views expressed by the author/publisher in this book do not necessarily represent the views of, nor should be attributed to, Sweetgrass Books. Sweetgrass Books is not responsible for the content of the author/publisher's work.

Printed on paper with 10% PCW content.

Printed in the United States.

15 14 13 12 11   1 2 3 4 5 6 7

Table of Contents

Preface   iv
Introduction   viii

Chapter 1. Climate Change: The Real Issues   1

Chapter 2. Earth's Oceans   25

Chapter 3. Earth's Temperature Balance   39

Chapter 4. Earth's History: a Short Course   71

Chapter 5. The Great Dying   95

Chapter 6. Earth's Carbon Balance   113

Chapter 7. Current Climate Change Issues   133

Chapter 8. What Can We Do?   151

FAQ: Answers to Frequently Asked Questions   175
Acknowledgments   185
References   186
Index   192
Illustration Credits   196

Preface:

While the bulk of this book describes Earth's climate through time, the premise of this work is that the Earth is a closed system, and ongoing processes involving earth, sea, and sky interact with one another. One cannot separate the depletion of atmospheric ozone from the effects of UV radiation increases on soils, for example. Yet, as humans scramble to understand and mitigate the effects of Earth's increasing temperatures, we tend to isolate and compartmentalize these processes and events. Like the old story of the blind men examining the elephant, hundreds of research labs and thousands of researchers are working on various aspects of Earth's changing climate and ecosystems worldwide, yet there is no concerted effort to understand the interactions and ultimate result of those interactions on Earth's future ecosystem.

This work provides a new approach to understanding what will happen in the "near" future, based on what, according to the rock and fossil records, has happened in the past. And, more importantly, this work offers insights into global climate interactions currently being overlooked or ignored because they don't "fit" in any particular scientific discipline's "compartment." Unfortunately, cataclysmic events of global impact have occurred and will again occur. In order to understand causes and timing of these major events, one needs to synthesize data from many scientific disciplines. Paleontology and paleontologists are eminently suited to this purpose.

Small events of short duration (geologically) such as ice ages can cause large changes in ocean depth, and thus increase or decrease terrestrial landmasses significantly. Yet, for the most part, life adapts relatively easily to these changes. It is only when catastrophic, cataclysmic events happen that the Earth loses most of its living organisms.

Giant-size extinction events happen rarely. The fossil record only records 4 (arguably 5, depending on at what level you draw the line) that caused that sort of calamity. However, as is shown in this work, we are poised to encounter the next.

Humans will likely survive the loss of half of Florida; humans will likely not survive the loss of half the Earth's atmospheric oxygen. The information presented is designed to help humans stave off the next of the "big 5 extinction" – type event. At best, the timings, magnitudes, and resultant catastrophe levels proposed in this book are overstated, but humans still react as suggested. The end result would be a cleaner, sustainable environment as the Earth's future biosphere. At second-worst (the worst is loss of virtually all life on Earth, humans included) the problem is understood and we, as humans, stave off the worst of the effects. In either case, the human species in particular and life on Earth in general benefits from understanding and following the procedures recommended in this book. However, to do nothing, or to do the wrong thing, could have disastrous consequences, the likes of which the human species has never experienced. We need to pay attention to what the Earth itself is telling us and to act appropriately. Any other course of action could lead to our grandchildren really, really hating us, if they survive at all.

Creation, Paleontology, and the Rock and Fossil Records

I'm sure you are wondering how paleontology and current climate change issues relate to one another. Some, I am sure, will also argue that this work is meaningless, since science has the past all wrong - it doesn't agree with scripture. I want to address this issue.

The first point I wish to make is that much of the rock and fossil record exists, according to scientific interpretation, due to repeated climate change. If climate and ecosystems hadn't changed repeatedly and significantly, we would mostly see only one type of sedimentary rock piled up throughout Earth's history in any given area. Instead, the presence/absence of polar ice caps caused sea levels to vary; variations in the amount of precipitation caused correlated variations in types of sediment deposition, and the occasional catastrophic event caused dramatic, geologically instantaneous changes in sediment content and deposition.

Modern environments allow us, as paleontologists, to match the sediment resulting from existing processes and conditions to those we observe in the geologic record. Likewise, those sediments directly above the ones that match Earth's current conditions most closely should be excellent predictors for what will happen in our future if Earth's events follow the same general patterns. And it is this pattern-matching and prediction process that provided the bulk of the data for this work.

The second point is that, coming from the basis of science, the above-mentioned patterns are assumed to be random, natural occurrences. If, however, the Earth is here by divine design instead, and the patterns are from intelligently designed purpose, that record suddenly becomes clear warning rather than happenstance. And at that point, the warning is directly from the hand of the divine.

Humans can argue forever about the mechanisms for the origin of the message presented in this book. However, the message recorded in the rock and fossil records is the same regardless of how it came to be originally provided or recorded. I would strongly urge those who may be offended by the "Earth history" dating used in this work to not allow such to interfere with understanding the message itself. The message is far too important, and warnings given should never be ignored.

Introduction

In my work as a dinosaur paleontologist, I have spent much of my life studying the Earth's rock and fossil records. The fossils themselves and the sediments in which they are buried often combine to tell a story of the climate, environment, and, sometimes, catastrophic events at that point in time. One of the first things I observed was that local, regional, and global climate factors played a large role in preserving any specimen. We were putting together snapshots of Earth's history, one instant and one location at a time; unfortunately, these snapshots are most often used to portray a generalized trend covering millions of years. Seldom does the public have access to the more detailed information.

With the advent of new equipment and techniques, the last few years have been marked by tremendous increase in the amount of data we can use and correlate. This increase in data volume represent increases in both the accuracy of interpretations and the level of detail available to us. The picture we now have available is one of an Earth that has supported life virtually as soon as liquid water appeared. This life was likely limited to simple forms that could survive tremendous variations in chemical composition and temperature of their habitat. It wasn't until the Earth's climate and chemistry stabilized that more complex life forms were able to survive and thrive.

Humans instinctively know that their lifestyle depends on a stable, predictable environment. Only since the retreat of the last glacial ice and the subsequent stabilizing of our environment has human civilization flourished.

Much of our current success can be attributed to this stable environment. However, at an unconscious level, most of us understand that this stability is fleeting. We worry about catastrophes and global environmental changes, and this worry is most likely an instinctive rather than cognitive reaction. Humans have survived times of unstable environment, but dealing with these changes has left its mark on humanity.

My work on dinosaurs has led me to a deeper understanding of the complex and long-term patterns of Earth's environment. So, when I hear on the news or read in the newspaper about the latest climate catastrophe caused by human activity, I have to put this in context with the catastrophes recorded in Earth's rocks and fossils. Then, at least most of the time, I have to shake my head at our anthropocentric view. I wonder: if dinosaurs were smart enough to understand, would they shake their heads in amazement at these recently evolved mammals that seem to think they control the world?

I have tried to write this book in a simple and readable form. However, you will find references cited on the various aspects of this issue should you wish to delve more deeply into a topic. Readers are encouraged to check my conclusions independently. While we scientists argue over the details, the general public is left uninformed or misinformed, and politicians are asked to create policies to preserve our way of life without fully understanding the problem. I have tried to explain the world carbon system and history at a level that is not so simplistic as to be misleading or inaccurate and yet is still understandable and not buried in the details and controversies. Because the system is so complex, any number of events could skew the timetable and/ or magnitude of the events described in this work. I have tried to suggest a "middle ground" approach to protect ourselves and our future while avoiding the extreme. I feel that, while it is much better to prepare for an event that turns out less severe than we expected than to face cataclysm, we must also live our lives. However, we must be aware of the problem and take the necessary steps to insure the survival of the human species. It would be a shame for our grandchildren to face annihilation because we were too busy living our lives or arguing over the details to solve the problem. I also feel we must be realistic in what we, as humans, can ultimately accomplish.

Not everyone will agree with my conclusions nor my suggested solutions, especially since they involve radical changes in our society and in our global energy budget and usage. However, I believe that, if the human race wants to save itself from impending disaster, it must act soon. There is little time to waste, and wading through years of scientific, political, and economic wrangling could prevent us from taking the steps necessary to insure our survival. At worst, my suggestions provide posterity with the means to sustainable balance in the energy we need without further damaging our ecosystem. At best, they provide the means of survival for our posterity. The price is merely giving up those activities that are not sustainable and that damage our ecosystem. I personally believe that this is a small price when balanced against the potential benefits.

As passionately as we humans ponder our existence on our planet, the ultimate question remains unanswered: Can the human race chart its own course in light of the climatic and catastrophic challenges ahead? I believe that, unless we include the information provided by the Earth itself concerning long-term cycles, we will never be able to answer that question.

Humans have not been on Earth long enough to directly observe even short-term cyclic events such as magnetic polar reversals, much less the long-term icehouse-greenhouse events that have occurred. Even when humans have observed climate-changing events such as the "mini Ice-Age" of the Middle Ages, we can't agree on the causes. We do know that short-term events can affect global weather patterns. One example is Mount Tambora's eruption in 1815 that led to the year 1816 being called "the year without a summer." How much more important it is, then, for us to look at events of longer cyclic periods and duration! The past is truly the key to our future.

Chapter One

Climate Change:
The Real Issues

A few years ago, the Intergovernmental Panel on Climate Change, IPCC, presented its findings. That report showed that human activities have affected Earth's global temperature balance, and a massive worldwide effort to mitigate these effects has been undertaken. The entire IPCC won the Nobel Prize for their contributions to our understanding of Earth's climate and human interactions. The focus of their work has been the potential disasters we face as a result of the Earth warming a few degrees.

However, not all scientists are convinced. Some argue that we are only seeing "natural" processes coincident with human activity. These scientists believe the Earth's systems are too complex and massive for a factor as small as human interactivity to affect. Others argue that

Earth's temperature balance is cyclic, and we are just observing never-before seen cycles. Still others argue that, regardless of the cause, a warmer Earth is a good thing. Disagreement as to causes, effects, and importance of these

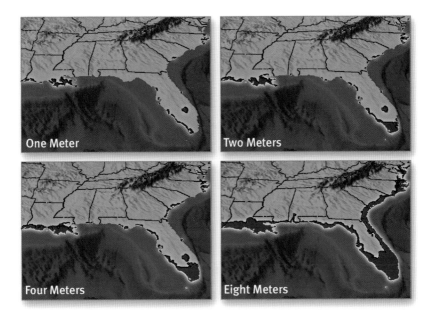

Figure 1. Map of Southeastern United States with inundated areas in red based on the listed sea level rise. (courtesy of NOAA)

global climate changes has led to an increased confusion and skepticism within the public's viewpoint.

My perspective, from a paleontological point of view, is centered more on the longer-term survival of the human species than it is with whether Miami, New Orleans, and Washington D. C. will be underwater in a few years (see Figure 1 with inundated areas in red based on the listed sea level rise, courtesy of NOAA).

From the sea level image, you can see that even if the ocean levels were to rise over 20 feet, the bulk of Earth's landmasses would remain above water, and they would still be able to provide food and living space for humans. In fact, what would be lost in Florida and the coastline would be more than made up for in new habitable and productive areas in Canada, Russia, and Greenland. While countless individuals will undoubtedly be affected, I am not convinced humans have any ability to regulate Earth's temperature balance so as to prevent such events. From a paleontologic perspective, such events are the norm.

The issues dealt with in this work concerning climate change are primarily focused on those events that could remove the human species entirely. And, from a paleontological perspective, those events, while not common, also happen on a regular basis. I am concerned that, if we exert our efforts in the wrong manner, we won't have the resources to protect our species from the truly devastating events the Earth experiences. In a landslide, one does not survive by landing under a boulder because he was dodging a pebble!

The rest of this chapter is devoted to a summary of some of the more important interactions between Earth's composition and processes, human activities, and global climate. These topics are designed as a review and evaluation of the concepts and misconceptions currently presented to the public concerning climate and climate change.

Earth's Oceans

Earth's oceans provide the bulk of the interactions that regulate climate on our planet. The entire next chapter is devoted to describing some of these relationships. However, because ocean/climate interactions are so integral to our understanding of our future ecosystem, I felt it appropriate to mention a few points here as well.

Climate change in the form of the current global warming trend will cause sea levels to rise. This rise will not be uniform due to the effects of the Earth's rotation, tides, and currents. Because most of Earth's current population lives

Figure 2. Beautiful as they are, coral reefs are among the most sensitive enviornments on earth to changes in climate.

at or near sea level such changes are perceived as a serious problem. In some places, extensive efforts have staved off the ocean's encroachment, and the Netherlands actually added significantly to its landmass by draining ocean-covered areas. As sea levels rise, it will become increasingly difficult to prevent disasters when unusual conditions present themselves. A prime example is that of the difficulty for humans to fend off the high waters driven by Hurricane Katrina. Rises in sea level, and human efforts to prevent ocean encroachment, will lead to increasingly more serious and widespread disasters when such efforts fail.

Another effect of rising sea levels is the potential loss of significant areas of reef structures. Reefs currently cover less than one percent of the ocean floor, yet these reefs support an estimated twenty-five percent of all marine life. Reef builders are, for the most part, quite sensitive to water depth, water temperature, salinity, pH, and chemical content. Currently, many of these factors are being affected by climate change. In most areas, coral reefs are in danger. That the current ocean pH, temperature, and chemistry shift is man-caused is well-documented (Dorritie, 2007; Hansen and Sato, 2004; Tans, 2009).

Earth's major ocean currents are driven by a process known as "thermohaline circulation." In a very simplified form, this process can be understood as follows: Cold, low salt content water sinks to the bottom and flows toward the equator from the Polar Regions. In turn, this water, as it travels, mixes with warmer and saltier water. Warmer waters rise and are further heated. Since the cold, polar waters have to be replaced, this leaves only the warmer equatorial waters with the ability to flow pole-ward and re-supply the Polar Regions. The shapes of Earth's various landmasses and their positions on the sphere, combined with the effects of Earth's rotation, the Moon's gravitation, and a host of other factors, causes the ocean currents and circulation patterns observed today.

These circulation patterns are transient and easily disrupted, geologically speaking. The easiest way to disrupt the circulation is to significantly decrease the amount of polar melt water flowing toward the equator. Such a decrease reduces the amount of force driving the circulation patterns, and the obvious result is a curtailing of thermohaline circulation. The volume of cold, low-salt waters flowing from the poles is greatly lessened, and thus they would not flow as far toward the equator before warming. Likewise, a similar reduction in warm water flowing toward the poles would result in less warm surface water flowing northward, causing less precipitation in northern latitudes and providing less energy to drive storms. Overall, this

Figure 3. Highly specialized animals such as the koala cannot survive rapid changes in their habitats.

leads to a drier and warmer northern climate regime. Since most terrestrial climate is driven by these circulation patterns, initially small changes in the thermohaline circulation can result in large climate changes.

Ocean chemistry is as important to life on Earth as ocean temperature. One of the effects of the CO_2 in the atmosphere is the global lowering of the pH of the ocean waters. This topic will be discussed in more detail later in this work. However, in short summary, one of the effects of this lowered pH is a massive die-off of pH-sensitive organisms. In turn, these organisms, mostly microscopic, form the basis of the marine food chain. Much higher up the food chain, a major portion of Earth's human population relies on fish that were, in turn, able to survive by eating food based on the previously mentioned microscopic organisms.

Another effect of the die-off of those microscopic marine organisms would be the oceans' increasing inability to remove CO_2 and other greenhouse gases from the atmosphere. As the oceans become more inhospitable to life due to chemistry and temperature changes, less greenhouse gases will be removed by ocean-based factors. In turn, this will cause more severe climate changes, again reinforcing the cycle to the point where the removal of CO_2 and other greenhouse gases is negligible. Earth's present climate and Earth's ability to maintain current life forms, at that point, would be seriously disrupted.

The real time bomb that is ticking away is the methane currently preserved in ice under the oceans and tundras. This topic will also be discussed at length later in this volume. The short message, however, is that, should that substance, methane clathrate (also known as methane hydrate or methane ice) be released in quantity, a global catastrophe would be unleashed rivaling the worst asteroid impact or solar flare. This ultimate result of global temperature rise would make it highly unlikely that human life could survive in any form.

Life on Earth is tied most closely with ocean conditions. In the paleontologic record, virtually any disturbance in ocean conditions resulted in accelerated extinction events, and major shifts corresponded to global catastrophes. Thus, based on fossil evidence, the environment we should be protecting, if we wish to remain a viable species, is the ocean.

Figure 4. Oil spill in Gulf Shores, Alabama. June 12, 2010.

Rainforests and Plant Life

Rainforest areas are shrinking due to the Earth's climate becoming warmer and drier. Added to this effect is the impact of current human activities. These factors combine to destroy between 14 and 32 million acres of tropical rainforest annually (Laurance, 2010). Over 20% of the Amazon Rainforest has already been cut down (calculations and estimates vary –

During times of rapid climate change such as those we are seeing today, the die-off portions of the cycle are favored.

the number presented is on the conservative side of those reported). This activity does affect both atmospheric gas exchange and Earth's albedo--its reflectivity. In Asia and Africa, the natural process of desert encroachment on areas currently covered by rainforest competes with human activities as the largest cause of rainforest loss (Sivakumar 2007). Together, the loss is significant and accelerating.

Rainforests are "mature" ecosystems consisting of thousands of mostly specialized species interacting in the precise manner that maintains a delicate balance. On the other hand, the farmlands and grasslands that are replacing the rainforests are home to only a few species. Since many of our medicines and other chemicals have been derived from rare plants and animals found in the rainforest, the success of the human species is tied, in part, to the success of these regions. Each species lost represents a potential lost interaction with humans as well.

The ultimate effects of rainforest loss on Earth's albedo, Earth's CO_2 balance, and Earth's atmospheric circulation, humidity, and precipitation patterns are unknown. Most of the cleared lands are planted back to crops, so the "green" covering is likely to prevent serious albedo changes. However, at this point human records only go back a few thousand years. Global effects of the very few large-scale changes to Earth's biosphere that occurred during that time have not been well-documented. We must rely on the fossil record to provide us with glimpses of such events. And, the fossil record suggests a much more violent and energetic Earth –one often affected by events of magnitudes not witnessed by humans. In the fossil record, such massive restructuring of Earth's plants are invariably linked to major animal extinctions as well. They are also often accompanied by major climate shifts.

It is safe to say that it is in the best interest of human populations that we don't experience such, and the best way to attempt to maintain the status quo concerning this lack of catastrophic events is to maintain the *status quo* of conditions currently present.

Human Population Dynamics

Present trends in human activities, coupled with the natural progression of climate change on Earth, paint a grim future picture for humans and other higher life forms. Initially, the Earth will continue to lose species for a number of reasons. Species that are so specialized (adapted to a very specific set of

Figure 5. Volcanoes can have dramatic short-term climate effects.

environmental conditions) that they cannot adapt are being lost as climate changes occur. Other species are being, and will continue to be, lost due to the natural processes of interspecific competition and predation. Some species are just naturally able to succeed at the expense of others, either by out-competing them for resources or by eating them.

Still other species are being lost to human environmental modifications. Every time we clear the land, pave a parking lot, or conduct a myriad of other activities, we affect the habitat of all organisms there. If those habitats are consistently in areas chosen for human interaction, entire species can be lost. While species displacement by other species has been a traditional driving force of evolution, there is no evidence that any species before humans has changed the global environment so profoundly in such a short time.

While these factors are disturbing, they account for no more than a small rise in extinction rates over what has happened in the past. Evolution and extinction most often follow a pattern of expansion, diversification, specialization, and extinction repeated over and over through time. At a time where there are abundant resources in a given area and limited life forms using those resources, other life forms move into the area to take advantage of the abundance. These forms often rapidly diversify due to their individual abilities to thrive on different things to eat, temperatures, and water availability. This resultant diversity is then reduced through competition. Some animals are naturally able to protect their resources better than their competitors, and the losers eventually vanish. Unfortunately, this specialization comes at the price of being unable to adapt when rapid change occurs in the environment, and the specialized forms die off, leaving resources with no one using them. The process then repeats.

During times of rapid climate change such as those we are seeing today, the die-off portions of the cycle are favored. Over the past 250,000 years, we have seen a tremendous turnover in the Earth's species due to the extreme climate changes of the glacial/interglacial cycles. In fact, these cycles are likely a huge factor in human evolution and success (Hetherington and Reid 2010). For the first time, a species exists on Earth that is capable of modifying its environment even in extreme conditions to match its needs, rather than matching its survivability to the conditions presented. This process has allowed humans to deal effectively as a single species with virtually all climate conditions on Earth's terrestrial surfaces without significant diversification. And, over the past few hundred years, this species has prospered.

Rapid climate change (as predicted) will have a profound effect on the human species' ability to provide for itself. In the past, we have had a combination of interspecies interactions and climate effects that have limited human population growth. This is no longer the case. Because we have had a few hundred years of relatively stable climate, coupled with the abilities gained in the industrial age, we, as a species, have been able to provide for many more individuals and have produced them. We have established a fragile balance between what it takes to support our species and what it takes to support the rest of life on Earth. However, even very small changes in climate, relative to what is recorded of Earth's history in the rock and fossil records, could seriously disrupt this balance.

Scientists intimately involved in global warming research have discounted all but the most recent of data, saying older data is irrelevant (Running, pers. comm., 2008). They insist that, because climate patterns were not the same as they are now, predictions and outcomes of those past patterns and events would produce different results from those we are now postulating based on the current conditions.

The rebuttal to this argument is at the heart of this work. Granted, the Earth's climate has changed, and the complex models created to understand today's climate would not necessarily apply to earlier points in Earth's history. However, the exact nature of the specific causes and effects that these models use is not important; it only matters that the past history of the Earth shows a cyclic, predictable pattern. Indeed, the past may be a better key to what is now happening, because only the most significant events are preserved in the rock and fossil records. The complexity observed in the modern system is thus greatly simplified in the records of the past, and the major factors are more easily recognized.

Does it matter whether the CO_2 is produced by humans burning stored fossil carbon or that CO_2 is injected into the atmosphere by erupting volcanoes? Probably not. The effects of the infusion of CO_2 are going to be similar. The collapse of the Earth's biosphere 251 million years ago during which the Earth lost an estimated 90% of its plant and animal species and half of its breathable oxygen (Kiehl and Shields 2005) was catastrophic at that time. It would be similarly catastrophic if it happened today. Predicting, preventing, and/or mitigating such an event is the most important purpose for studying Earth's past.

Most often however, paleontology is seen as a "pure" science with no applied value. Most people do not recognize that one of the goals (arguably the most important one) of paleontological research is the gathering of data to create models applicable to the present and the future. Paleontology is a synthesis of many branches of science – geology, ecology, biology, forensics, and geochemistry to name a few – to create a more accurate understanding of prehistoric organisms and their environments. Dinosaur paleontologists like myself take this research and apply it to the only extinct group of animals to rival the diversity and complexity of the modern mammal-dominated ecosystem.

But how does all this work? How do we apply to the present what we've learned from the past? And, most important, why should we believe that what we are observing from the fossil, rock, and ice records have any bearing on what's happening today?

To answer the first question, we must invoke statistics and the law of averages. The Earth is obviously an extremely complicated place where each organism, cloud, or rock outcrop can have an influence on the overall climate balance of the biosphere. However, a multiplicity of influences over a period of time will balance each other and will produce an average series of conditions that is stable and predictable.

For instance, if you had a sack of marbles of various colors, as a single event you could draw out any one of the represented colors. However, if you shake up the bag and draw out a marble, and do it over and over again, eventually the numbers of each color drawn would closely match the actual

A more important aspect of dinosaur research . . . is how the dinosaurs came to rule the Earth in the first place.

numbers of the various colored marbles in the sack. Thus the unpredictable effect of a single draw becomes very predictable when combining the effects over time. Likewise, because there are literally millions of interactions going on, the effect of one climate factor is almost always balanced by the others, and climate is very predictable over time.

We will discuss the major factors affecting global climate later in this book. However, only two of those major influences have been shown to regularly shift Earth's temperature equilibrium. First, changes in atmospheric composition produce almost immediate corresponding climate effects. There are processes associated with living organisms that gradually change the composition of the Earth's atmosphere and biosphere. Such processes are responsible for converting our primitive atmosphere into one containing free oxygen, creating huge deposits of limestone worldwide, and cooling our planet by converting carbon dioxide into coal and other hydrocarbon compounds.

Second, Earth's global average temperature may be disturbed when some major event, such as a major volcanic eruption or extraterrestrial impact, occurs. Usually however, these events are of short duration, at least on a geological scale. Long-term changes usually require repeated events such as serial volcanic eruptions over a period of thousands or millions of years. A single event may cause initial climate change, but unless the resulting change in the physics or chemistry of the biosphere is repeated, the Earth's biosystem rapidly returns to its equilibrium state.

There are only three variable factors that, from the entire geologic record, are shown to have the ability to interact to regulate and vary Earth's climate over the long term. These are the amount of energy reaching the Earth from the Sun, the reflectivity of the Earth's surface (its albedo), and the composition of the Earth's atmosphere.

The terms, "energy capture" or "energy entrapment" are commonly used to describe any and/or all of the processes whereby solar energy is converted, passed along, or used, by various mechanisms before being released back into space. Because other sources of energy that affect Earth's temperature are stable – i.e. the heat escaping from the Earth's interior – variations in solar energy capture and global climate change are, in practicality, synonymous.

As can be seen in Figure 6, all the energy from the Sun that hits the Earth eventually makes its way back into outer space. However, the energy that is simply reflected has no noticeable effect on Earth's climate. On the other hand, the energy absorbed by the land, atmosphere, and oceans causes the Earth to remain warm and habitable before it is, in turn, lost to space and replaced by other packets of energy from the Sun.

From the illustration in Figure 6, it is obvious that eventually the energy incoming from the Sun is balanced by the energy reflected and radiated. However, as the Sun's energy is transferred through the various processes, its radiation back into space is delayed. These delays can cause small imbalances in any one or more of the processes indicated by the diagram, and thus can raise or lower the temperature of the land and oceans considerably. For instance, this excerpt from Kious and Tilling, 1996, describes the effects on global temperatures from 2 volcanic eruptions:

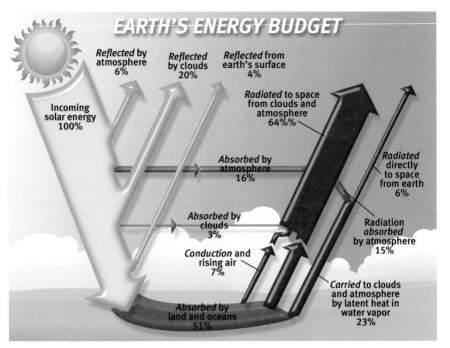

Figure 6. Earth's energy budget.

". . . the impact of the June 1991 eruption of Mount Pinatubo was global. Slightly cooler than usual temperatures were recorded worldwide and the brilliant sunsets and sunrises have been attributed to this eruption that sent fine ash and gases high into the stratosphere, forming a large volcanic cloud that drifted around the world. The sulfur dioxide (SO_2) in this cloud -- about 22 million tons -- combined with water to form droplets of sulfuric acid, blocking some of the sunlight from reaching the Earth and thereby cooling temperatures in some regions by as much as 0.5 degrees C. An eruption the size of Mount Pinatubo could affect the weather for a few years."

A similar phenomenon occurred in April of 1815 with the cataclysmic eruption of Tambora Volcano in Indonesia, the most powerful eruption in recorded history. Tambora's volcanic cloud lowered global temperatures by as much as 3 degrees Celsius. Even a year after the eruption, most of the northern hemisphere experienced sharply cooler temperatures during the summer months. In parts of Europe and in North America, 1816 was known as "the year without a summer."

Two points can be made from these observations. First, volcanic eruptions can cause the Earth to cool significantly and rapidly. And second, the effects do not last. Even though the Tambora eruption lowered the global temperature nearly 3 degrees Celsius (over 5 degrees Fahrenheit), this effect was very short-lived. In order for the Earth's temperature balance to be affected over the long term, multiple eruptions would have to occur. It should be noted that most large meteor impacts follow this pattern as well.

Paleontologists have been studying the rise and fall of the dinosaurs for over 100 years. Lately, much attention has been given to the Cretaceous-Tertiary extinction, when all dinosaurs with the exception of birds disappeared. Unfortunately, the disappearance of the dinosaurs has, primarily by popular press, been tied to a global catastrophe caused by an asteroid impact. In reality, a number of events were occurring simultaneously, and each undoubtedly contributed its share to the demise of the dinosaurs (Prothero 2009). However, since so much press has been given to the asteroid impact, its actual contribution to the demise of the dinosaurs has likely been overrated.

A more important aspect of dinosaur research, as far as its bearing on modern global warming issues are concerned, is how the dinosaurs came to rule the Earth in the first place. We now know that a catastrophic event that wiped out over 90% of life on Earth allowed the dinosaurs' great-great. . . grandparents to survive when most other forms of life perished. This event, known as the Permian/Triassic extinction, is the single worst catastrophe experienced by complex life on Earth.

At the end of the Permian Period, a glaciation event occurred that was similar to the one we experienced over the last million years. This event was followed by a slow warming of approximately 5 degrees Celsius (9 degrees Fahrenheit) over a period of tens of thousands of years. Then, a geologically instantaneous warming event (speculated by some to have taken no more than 3 to 15 years) (Dorritie 2007) raised Earth's average temperature an additional 6 degrees Celsius (10 degrees Fahrenheit) or more on top of the previous warming (Kearsey et al 2009). It is the study of this event and its parallels with the Earth's current conditions that place paleontologists in an ideal position for understanding the current climate changes and predicting future events.

The actual contribution of the asteroid impact to the demise of the dinosaurs has likely been overrated.

Chapter Two

Earth's Oceans

A Powerful Effect

Earth has been referred to as the "water planet" because, of all the planets in the solar system, only Earth is mostly covered by liquid water. This water surface provides the vast majority of interactions that stabilize and regulate Earth's climate. Earth's oceans trap the majority of the solar energy that strikes our planet - as much as 98% in regions and times where the sun's rays and the water surface are within 30 degrees of perpendicular (Jin, et al. 2004). The ocean currents then redistribute this energy, allowing areas of the planet that would normally be very cold to be significantly warmer. This interaction between the oceans and the Earth's temperature balance is by far the most important factor in equalizing global temperatures.

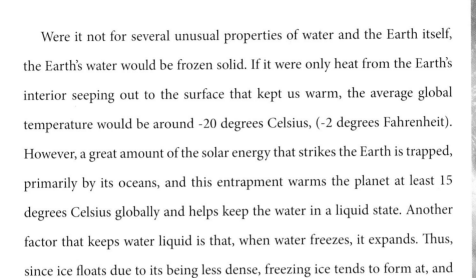

Were it not for several unusual properties of water and the Earth itself, the Earth's water would be frozen solid. If it were only heat from the Earth's interior seeping out to the surface that kept us warm, the average global temperature would be around -20 degrees Celsius, (-2 degrees Fahrenheit). However, a great amount of the solar energy that strikes the Earth is trapped, primarily by its oceans, and this entrapment warms the planet at least 15 degrees Celsius globally and helps keep the water in a liquid state. Another factor that keeps water liquid is that, when water freezes, it expands. Thus, since ice floats due to its being less dense, freezing ice tends to form at, and float to, the surface. Freezing, therefore, almost always occurs from the top down. This leaves the ability for warmed, tropical waters to still flow toward the poles beneath the ice, even where/when ice covers its surface, and allows relatively rapid global energy distribution. In turn, this energy distribution system tends to give rise to much warmer mid- to high latitudes than would otherwise be the case.

The Earth's oceans provide for a stabilization of temperatures on Earth between day and night, between high and low latitudes, and through fluctuations in the Sun's energy output. The fact that it takes a LOT of heat to change the temperature of water, coupled with the Earth's large water content and water circulation, means that huge changes in the energy being received from the Sun from the factors mentioned above result in relatively small variations in the Earth's climate. The oceans act as a huge energy reservoir, providing heat to the system when solar output is low, storing heat energy when the output is high, and circulating it so that its distribution is more uniform.

A great amount of the solar energy that strikes the earth is trapped, primarily by its oceans.

Evaporation takes energy. As solar energy warms the ocean, evaporation carries energy higher into the atmosphere than does the direct energy transfer of heat from the ocean surface. This is especially true near the equator, where warm air and moisture are carried high above the Earth and then drifts toward the poles. By the time this heated air mass has traveled some 1500 miles (2400 kilometers) northward, the heat energy is slowly dissipated, and the cooling air sinks lower and closer to the Earth's surface. At some point, this air is no longer being cooled from the loss of heat, but is once again being heated from the Earth's surface. However, by then most of the water vapor carried in this air mass has condensed and precipitated, leaving this final stage very dry as it approaches the Earth's surface. For this reason most of Earth's deserts occur between 20 and 30 degrees latitude, both North and South. This particular atmospheric phenomenon is called the Hadley Circulation.

In a simplified view of these atmospheric circulation patterns, the most intense and dependable flows happen from the Equator to approximately 30 degrees latitude, a second set from 30 to 60 degrees latitude, and a final pattern from 60 degrees to the poles. Only the equatorial flows are referred to as Hadley Cells. The other two are referred to as the mid-latitude cells and polar cells, respectively. As shown in the illustration below, these patterns interact to pump warm air poleward at altitude and cooler surface air toward the Equator, and this process is driven by heat transferred to the atmosphere from Earth's oceans. This generalized model illustrates the intimate link between Earth's oceans and its atmosphere and illustrates another of the major mechanisms by which the heat gathered by equatorial ocean waters is transferred toward the Polar Regions.

Evaporation and precipitation's cycling of water vapor through the atmosphere also removes atmospheric impurities. Water vapor condenses

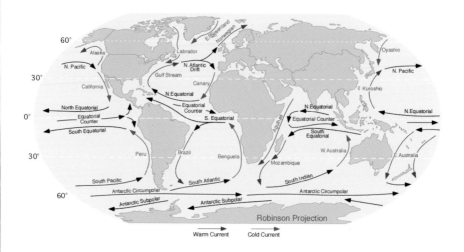

Figure 7. A map showing the primary ocean currents, the "Hadley Circulation".

around these small particles, and then carries them back to the Earth's surface once the droplets become too large for the air to support. Were it not for this cycling, ash particles from volcanoes, ash particles from both natural and manmade fires, fine dust stirred up from the Earth's surface, and a myriad of chemicals and other substances would remain in atmospheric suspension far longer. If that were the case, these substances would then form cloud layers that would remain in Earth's atmosphere indefinitely - much like those on Venus today. However, because atmospheric mist attaches to these substances and carries them back to Earth's surface, our atmosphere remains relatively clear.

The climate effects of ocean currents and circulation patterns are easily demonstrated. Warm surface waters carried by ocean currents keep Murmansk, Russia an ice-free seaport even though it is farther north than most of Alaska. Conversely, cold water flowing from the Arctic Ocean past the Bering Strait mixes with warm water of the North Pacific Current to form a cool, oxygen and nutrient rich water body called the Alaska Current. These waters are responsible for growing some of the world's largest food fish and crustaceans and for keeping northwestern North America's climate cool and moist.

In the Southern Hemisphere, there are no land masses to break up the water flow patterns in the mid to high latitudes, so atmospheric and ocean circulation patterns are fairly uniform from west to east. This pattern does not lend itself to the transport of energy to the higher Southern latitudes, especially in the atmosphere. Only the mixing of ocean waters carries significant heat energy further poleward. Thus, the Antarctic remains colder and drier than the Arctic.

The Earth owes many other life-maintaining effects to ocean activities as well. For instance, Earth's atmospheric oxygen content is regulated largely by marine processes. The oceans remove CO_2 and increase atmospheric oxygen content because they teem with photosynthesizing life, and this life content is nourished by breaking down carbon dioxide into carbon compounds and freeing oxygen molecules. While it is well-known that photosynthesis

Figure 8. A microscopic view of Algae dividing. These simple organisms account for most of the CO_2 removal from, and free O_2 added to, Earth's atmosphere.

breathe, what is not well-known is that produces the oxygen we breathe, what is not well-known is that it is the phytoplankton – the microscopic photo-synthesizing organisms in the Earth's oceans – that perform the bulk of this oxygen production. Most commonly, people are under the impression that trees are the major photosynthesizers when in reality marine organisms at the opposite end of the size scale perform this task! A quick and easy way to verify this for yourself is to look at satellite photos of the Earth. Although, in general, Earth's landmasses appear brown and Earth's oceans appear blue, if you look closely you will see large greenish sections. This color, visible from space, is produced by chloroplasts in photosynthesizing organisms. What is amazing is that there is as much or more "green" per square mile in the ocean as there is on land, and 80% of Earth's surface is covered by oceans.

The algal blooms seen in Figures 9 and 10 are tied to carbon dioxide levels in the waters as well as the water temperature. The amount of phytoplankton in Earth's oceans has been steadily rising with the increase in carbon dioxide available from the burning of fossil carbon fuels. While these blooms do produce tremendous amounts of oxygen to add back into the biosphere – nearly enough to replace that being consumed by humans' burning of fossil fuels - many species producing these blooms are harmful or deadly to other life. Red tides are becoming an ever larger problem, poisoning fisheries (and humans who swim) where these algal blooms occur.

The Earth's oceans are responsible for most of the conversion of atmospheric carbon dioxide to other carbon compounds. Carbon dioxide is removed from the atmosphere primarily by its inclusion within rain droplets, which end up mostly falling into the ocean. Marine photosynthetic organisms then use the carbon dioxide to create the hydrocarbon compounds from which their bodies are made and giving off oxygen as a waste product. These hydrocarbon bodies, once the organism perishes, either sink to the bottom of the ocean or are carried through the food chain. The bulk of the carbon in this form that eventually reaches the ocean floor actually falls to the bottom in the form of feces (Honjo 1997).

However, it should be noted that the majority of the carbon remains in the upper portions of the oceans, traveling through the food chain. Some of this carbon does make its way back into the atmosphere as animal-exhaled

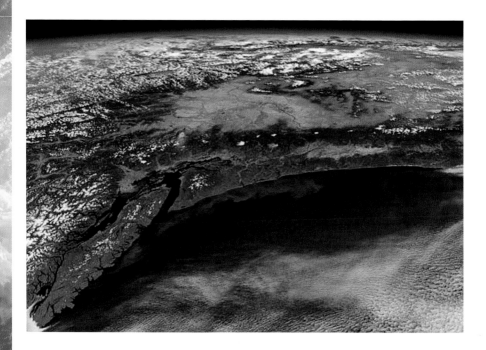

Figure 9. Pacific coast of North America, looking from west to east. Note the volume of green color in the ocean is nearly equal to the green on shore, and this shoreline is one of the greenest terrestrial localities on the planet. (Image courtesy of NASA)

CO_2. While the oceans do remove the excess carbon from the system eventually, it is a slow process and a delicate balance. Carbon dioxide trapped in water forms carbonic acid, a weak acid present in most soft drinks. In just the past few years, the oceans have become more acidic due to the addition of CO_2 from the burning of fossil fuels subsequently being carried into the atmosphere and then precipitated into the oceans (Wingenter et al 2007). If this process continues, the oceans will lose their ability to process CO_2 due to die-off of the photosynthesizing, pH-sensitive organisms that are integral to this process. That, in turn, will cause a cascade-effect that will lead to sterile, highly acidic oceans and hot atmospheric temperatures.

Figure 10. Algal bloom off the northern coast of Norway, deep in the Arctic Ocean. This picture is an apt comparison to the relative amounts of "green," indicating oxygen production on terrestrial versus marine surfaces around the world. (Image courtesy of NASA)

There is good evidence that most of the major extinction events of the past have occurred due to catastrophic changes in ocean conditions. In point of fact, most preserved sediments of the fossil record are marine sediments. Thus, the record is much more complete and detailed concerning what happened in the oceans long ago than that of what happened on land.

The chemical makeup of the oceans affects the chemistry of the sediments preserved on the ocean floor. Also, the sedimentary record from the ocean floor is much more complete and uniform than that occurring on land since sedimentation continues there at a fairly regular rate and there is little erosion. By examining these sediments, we can determine many aspects of the ocean chemistry at any given time in the past 4 billion years of Earth's history.

The chemistry of Earth's oceans has remained relatively stable through much of Earth's history. At times of rapid change in this chemistry, there are corresponding losses of fossil taxa – by definition, extinction events. For example, the most extreme extinction event known, the one that occurred at the end of the Permian Period, is correlated with a serious change in ocean chemistry. The following quote cites heightened hydrogen sulfide and carbon dioxide levels on the ocean floor as the major cause of this extinction event:

"New paleobiological studies on the environmental distribution and ecological importance of brachiopods, benthic molluscs, and bryozoans support the hypothesis that stressful ocean conditions—primarily elevated H_2S levels (euxinia) but also heightened CO_2 concentrations—were the prime causes of the end-Permian mass extinction." (Bottjer, et al., 2008)

While this is an oversimplification, the point is well-taken. Because the oceans house most of the life on Earth, and because the oceans are so intimately tied to global climate and ecology, what happens to Earth's oceans happens to the Earth in general.

What happens to Earth's oceans happens to the Earth in general.

Chapter Three

Earth's Temperature Balance

Perhaps the most difficult part of understanding what is real and what is fiction in discussions and reports on global warming is understanding the various components of the problem and their interactions with other factors. Indeed, the major controversies in current scientific debates revolve around which factors are responsible for which observations.

As mentioned earlier, only three factors have been consistently shown to affect Earth's global temperature balance – solar energy available, Earth's albedo (reflectivity) and Earth's atmospheric composition. However, the interactions between these factors is complex, and their influence varies from day to night, from location to location, from summer to winter, from land to sea, and from and to a multitude of other factors. Current climate research has focused on creating a computer model that can analyze all these factors and predict weather and climate more accurately.

Unfortunately, Earth's global climate system is so complex that no single model of causes and effects yet derived accurately portrays all of the observed data or is able to accurately predict in detail. However, there are observed phenomena that have unequivocally been shown to influence global temperatures as a general observation, and major patterns involving these phenomena are fairly accurately understood. If one models only the major interactions, these phenomena become understandable and predictable. This chapter examines a few of these major phenomena and their climate interactions.

Atmospheric Greenhouse Gas Content

Today's news is full of references to carbon dioxide, greenhouse gases, and global warming. Most of these references are used, in one way or another, to show how human activities affect Earth's climate. However, many of these claims are conflicting, and virtually none report on the percentages of greenhouse gases that are naturally occurring or their percentage comparisons with those produced through human interaction. In turn, these omissions have led some researchers (and a large portion of the public!) to skepticism concerning predictions of climate change. There is an old cliché concerning "throwing the baby out with the bath water," and I see this happening with skepticism concerning current climate data. In answer to this problem, the current top greenhouse gases are presented here in order of importance to the current and future entrapment of solar energy.

1. Water and water vapor

Water vapor accounts for over 60% of atmospheric energy entrapment (Trenberth and Stepaniak 2004). This is the pure, colorless, invisible gas – not atmospheric water in the form of clouds. Water is largely excluded from discussions of global warming since its presence in the system is self-regulating and not readily subjected to human modification. This is unfortunate, as the exclusion leads to inflated statistics concerning the actual influence levels of other greenhouse gases and their effects on global energy/temperatures.

The major factor regulating the amount of water vapor in the atmosphere is the temperature of the atmosphere itself. Relative humidity is a concept that most are familiar with, as it is reported as part of most weather reports. Absolute humidity is not often discussed. However, it is important to understand that, under the right conditions, an atmosphere only a few degrees warmer can hold several times the actual amount of water vapor and still have the same relative humidity. And, in the atmosphere, each added molecule of water vapor adds to the atmosphere's solar energy entrapment.

While water vapor is the primary factor in solar energy entrapment, the energy-gathering by water vapor occurs almost entirely in the lowest level of the atmosphere (the troposphere). Below-freezing temperatures only allow traces of water vapor to be present at the higher altitudes, and thus its effect at higher altitudes is negligible.

In addition to the invisible atmospheric water content, visible water also influences Earth's global climate. Water in the form of clouds and water and ice on the Earth's surface have entirely separate effects on global climate from that of the water vapor described previously. Water has characteristics that allow it to either be the most energy-absorbent or energy-reflective substance commonly found on Earth. The amount of water's energy absorption or reflection depends on the state in which the water is existing when bombarded. The largest single component of Earth's solar energy entrapment occurs through energy absorption by the oceans. However, solar energy is almost entirely reflected if the energy strikes the water at an angle greater than approximately 45 degrees. Polar ice caps and winter snows significantly increase reflectivity and thus effectively limit energy entrapment on Earth as well (Trenberth and Stepaniak 2004).

Clouds are by far the most complex of the water-related greenhouse factors. High, wispy, cirrus clouds readily trap energy, yet much solar energy is reflected back into space off the tops of lower-level clouds. Low-lying clouds that contain significant amounts of carbon dioxide, nitrogen oxides, and ozone are often referred to as photochemical smog, and these clouds are effective solar energy absorbers. Clouds also transport captured energy, absorbing the energy in one location and releasing it through precipitation in another. Clouds often force differences in reflectivity through their movements and precipitation, and these gradients may either enhance or limit energy entrapment, depending on the circumstances.

2. Carbon dioxide

Carbon dioxide is by far the most publicized of the greenhouse gases. It accounts for as much as 20% of the total energy trapped and 80% of the energy trapped in the upper atmosphere. In the past 100 years, the amount of carbon dioxide in the atmosphere has risen approximately 20%, from 290 parts per million to 369 parts per million. Most of the increase is from human sources – coal fired electrical plants, automobile emissions, and building heating. However, 10% of the carbon dioxide added to the system comes from burning timber and deforestation (Jacobson 2004). Also, large volcanic eruptions such as the relatively recent Pinatubo and Mount Saint Helens eruptions add significantly to the amount of atmospheric carbon dioxide.

Carbon dioxide is a stable gas. It can remain in the atmosphere indefinitely. This factor is the main reason it is ranked as such a serious problem. It is removed from the atmosphere primarily by photosynthetic plants. These plants convert carbon dioxide and water to free oxygen and complex hydrocarbons. However, this conversion reverses itself as the plants die and decay. Its effect on the system is neutralized only when the carbon in these plants is buried or otherwise removed from interacting in the biosphere.

In ancient times, carbon dioxide was much more abundant in Earth's atmosphere. For instance, it has been calculated that there was 20 times the amount of carbon dioxide in the atmosphere 500 million years ago (Berner and Kothavala 2001). Carbon dioxide levels 200 million years ago were still over 10 times present levels, and they were at 3 times the current level at the time of the dinosaurs' demise 64.5 million years ago. The past 20 million years or so have seen the lowest levels of atmospheric carbon dioxide present at any time in Earth's history, so far as we have been able to measure. And this lack of CO_2 in our atmosphere is matching one of the coldest periods in Earth's history as well.

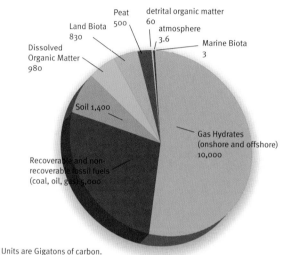

Units are Gigatons of carbon.
This graph is for Organic Carbon only. If dispersed Organic Carbon (kerogen, bitumen) were plotted, amounts would be three orders of magnitude larger.

Figure 11. Earth's Carbon reservoirs

3. Methane

Methane, otherwise known as natural gas or swamp gas, accounts for approximately 10% of the energy trapped by the atmosphere. It is over 20 times more effective than carbon dioxide in trapping solar energy (Schmidt 2004). It is the simplest of the hydrocarbon molecules, composed of a single carbon atom surrounded by four hydrogen atoms, and this structure is the basic building block of most organic molecules.

The importance of methane as a greenhouse gas would undoubtedly be ranked higher if it persisted in the atmosphere for a longer period of time. However, a typical methane molecule only lasts for an average of 10 years in Earth's atmosphere before it breaks down. The most common reaction by which methane leaves the atmosphere is oxidation. Methane combines with free oxygen to produce carbon dioxide and water. Unfortunately, this break-down produces CO_2 molecule-for-molecule and releases heat. Thus, methane released into the atmosphere contributes to global warming initially by trapping great amounts of solar energy, in an intermediate stage by burning and giving off a great deal of heat, and finally by becoming carbon dioxide and thus increasing solar energy entrapment indefinitely. Were the last two factors included in greenhouse calculations, methane would undoubtedly be ranked much higher in the greenhouse gas list.

The amount of methane in the atmosphere has nearly tripled in the past 150 years (CCSP 2006). Were it not for the fact that concentrations were very low to begin with, this statistic alone would be cause for alarm. However, the real cause for alarm is the fact that most of Earth's sequestered carbon resources involve this substance. Methane exists in large quantities as natural gas in landfills, coal beds, and in gas pockets in the Earth itself. Its nearly universal presence, ease of collection, and perceived harmless by-products make it the most desired source of energy for heating homes and businesses and for cooking worldwide.

Currently, over 50% of Earth's available carbon resources are tied up as methane (Buffett and Archer 2004). Most of this trapped methane is in the form of methane clathrate, also known as methane hydrate or methane ice. It is found on the ocean floor and under the tundra. In the ocean, it is commonly found within the mud of the ocean floor at depths of 200 to 500 meters (650 to 1600 feet), although clathrates have been observed at depths up to 2000 meters (6500 feet).

Methane clathrate is a crystalline solid that looks like ice, and that occurs when water molecules form a cage-like structure around the smaller methane molecules. Clathrates are also called gas hydrates. Gas hydrates were discovered in 1810 by Sir Humphrey Davy and were considered to be a laboratory curiosity. In the 1930s clathrate formation turned out to be a major problem, clogging pipelines during transportation of natural gas under cold conditions. Methane clathrate played a major role in the oil platform explosion and subsequent oil leak in the Gulf of Mexico on April 20, 2010.

Methane clathrate is formed by the decomposition of organic matter on the ocean floor and by subterranean methane venting that allow methane gas to bubble up into a dense, pressurized, cold water layer. Clathrates form through a unique crystallization of water into ice. Normal ice is crystallized in the hexagonal system, while clathrates are formed in the cubic system. In this structure the "cages" formed by the frozen water are arranged in such a way as to trap other gases. If all cages in its structure would be occupied by methane, one cubic meter of solid clathrate could contain 170.7 m3 of methane gas at standard conditions of temperature and pressure. In nature, one cubic meter of clathrate turns out to contain up to 164 m3 of methane (Chatti et al 2005). As seen in Figure 11, most of the carbon on Earth is actually trapped in this form.

Figure 12. Mexico City shrouded in photochemical smog.

A second effect associated with methane clathrate is that as it forms on the ocean floor, it creates a layer that is impermeable to gases. This impermeable layer on the oceans' continental shelves has further trapped methane gas underneath. Thus, not only is there a significant reserve of methane gas trapped in the form of clathrate, there is perhaps as much methane trapped as compressed natural gas beneath the clathrate layers.

The ramifications of this large, unstable methane reservoir will be further discussed in Chapter 5.

4. Ozone

Ozone is a highly reactive, unstable form of oxygen. Its chemical formula is O_3, instead of oxygen's typical O_2 molecular structure. It is both a necessary component of Earth's atmosphere and an effective greenhouse gas. In the upper atmosphere, ozone is formed when certain wavelengths of ultraviolet radiation strike oxygen molecules, splitting them into their constituent atoms. These atoms use energy absorbed from other wavelengths within the ultraviolet spectrum to bond to other oxygen molecules, thus forming ozone. Still other ultraviolet wavelengths break down the O3 molecule, and the process repeats.

Ozone occurs naturally in a layer of very dry air around 15 to 35 kilometers (9 to 22 miles) above the surface of the Earth, but is relatively rare. If all the ozone in Earth's upper atmosphere were brought down to the Earth's surface, it would only form a layer 3 millimeters (approximately 1/8 of an inch) thick. However, this small amount of ozone provides a great service for life on Earth. Ozone in the Earth's upper atmosphere is responsible for either absorbing or scattering virtually all wavelengths of ultraviolet radiation. This feature makes ozone an important factor in the prevention of damage to living organisms by UV radiation.

Ozone in the troposphere, on the other hand, is both a health hazard and a greenhouse gas (Kiehl and Trenberth 1997). Its formation is primarily driven by incomplete combustion of automotive and industrial fuels and by release into the atmosphere of chemicals such as xylene. As such, the production of ozone in the troposphere is largely human-caused. Lightning is also responsible for ozone creation in the troposphere, but although significant amounts are produced when lightning strikes, this ozone tends to disperse. However, human-created ozone tends to be created in more stable environments, allowing the ozone to accumulate and become a primary component of photochemical smog.

Ozone's reactive nature allows it to be used to sterilize bottled beverages, neutralize cyanides, and disinfect hospital laundry. However, as a component of photochemical smog, it irritates the mucous membranes of the eyes, nose, lungs, and throat; it can trigger asthma attacks, and within the human body it can produce the free radicals responsible for most age-related organ damage.

As a major component of photochemical smog, ozone traps a significant amount of solar energy. It has only been during the past 50 to 100 years that such a low-lying haze has been common over major portions of the Earth. Indeed, it was only in the 1950's that the term "photochemical smog" was coined. Photochemical smog can raise daytime temperatures by as much as 3 to 5 degrees Celsius (5 to 9 degrees Fahrenheit). Conversely, during times of darkness it can insulate the Earth's surface so that energy captured during the daytime can't radiate back into space at night.

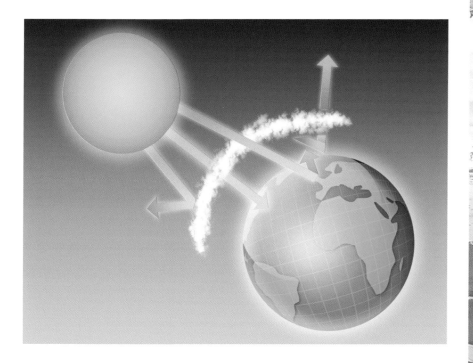

Figure 13. Ozone in the Earth's upper atmosphere is responsible for either absorbing or scattering virtually all wavelengths of ultraviolet radiation.

5. Nitrous oxide

This gas is listed fifth in this section, based on its importance in solar energy regulation. However, it is ranked third in its effect on global warming by the IPCC. The reason for this is that many scientists ignore two of the above-listed atmospheric components, water vapor and ozone. It is understandable that water vapor is left off the list, since human interactions have little to do with its presence or concentrations. However, why ozone is left off the list is a mystery.

Nitrous oxide is a stable molecule that can remain in the atmosphere for an average of 120 years. Because of this longevity and the fact that it is much better than CO_2 at trapping solar energy, a molecule of nitrous oxide has approximately 300 times the greenhouse potential as a molecule of carbon dioxide. While nitrous oxide is part of the reaction primarily responsible for creating ozone in the troposphere, it attacks and destroys the ozone in the stratosphere. Its concentration in Earth's upper atmosphere has increased from about 270 parts per billion in 1750 to 314 parts per billion in 1998, an increase of approximately 16% (IPCC Web site). Concentrations continue to increase at a rate of approximately .25% per year, or an extra 2/3 part per billion per year.

Two-thirds of the increase in atmospheric nitrous oxide is thought to be the result of human activities. While rotting vegetation provides a significant percentage of atmospheric nitrous oxide, higher percentages are produced by agricultural fertilization and by passing hot, burned fuel products through catalytic converters. By far the largest percentage comes from agricultural practices. Plants need nitrogen-based fertilizers, and some plants even produce their own through a symbiosis with nitrogen-fixing bacteria. Farmers also provide nitrogen fertilizer to their plants, and the least expensive is anhydrous ammonia, a gas that easily escapes its shallow burial and is easily broken down to form nitrous oxide.

6. Chlorofluorocarbons and other complex hydrocarbon gases

There are literally hundreds of combinations of carbon, hydrogen, sulfur,

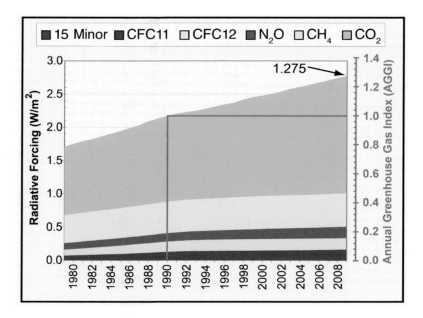

Fig 14. Radiative forcing, relative to 1750, of all the long-lived greenhouse gases. The NOAA Annual Greenhouse Gas Index (AGGI), which is indexed to 1 for the year 1990, is shown on the right axis.

chlorine, bromine, and/or fluorine that produce gases with energy-trapping potential. According to the Kyoto Treaty, these are grouped into 3 major categories - sulfur hexafluoride, hydrofluorocarbons, and perfluorocarbons. Of these, all but sulfur hexafluoride are carbon-based compounds.

The origin of these chemicals is primarily anthropogenic (human-created). Together, these gases account for about 10% of the estimated total greenhouse effect, excluding water vapor. As can be seen in Figure 14, carbon dioxide, methane, and nitrous oxide account for the other 90%.

These manmade compounds are relatively stable, and they all hinder the radiation of solar energy back into space. Perhaps their worst characteristic, however, is that when they break down, some of their constituent parts cause the depletion of ozone in the upper atmosphere. It was this factor that brought about a globally-enforced limitation on the production and use of certain chlorinated fluorocarbon gases (also known as CFC's or Freon). Contrary to popular press, the CFC molecules themselves do not deplete the stratospheric ozone. However, small, lightweight molecules produced as the CFC molecules break down (primarily Chlorine monoxide) do destroy ozone molecules. Bromine monoxide, produced by the breakdown of various halon gases, also causes catalytic breakdown of stratospheric ozone. The breakdown reactions for the large CFC and PFC molecules are precipitated in turn by ultraviolet radiation.

These complex hydrocarbons have been the focus of long-term production limitations, and polar levels of ozone show these limitations are working. Unfortunately, it has recently been shown that levels of some of these chemicals in the atmosphere surrounding urban areas is climbing, and that stratospheric ozone above major urban areas is dwindling as a result (Hopkin 2007). Such studies attest to the delicate balance achieved between the components of Earth's atmosphere and how easy it is to cause an imbalance.

Albedo - the Reflectivity of the Earth

One of the most important factors in Earth's temperature balance is its reflectivity. Fresh snow can reflect up to 95% of the solar energy that strikes it. In contrast, Earth's oceans typically reflect approximately 10%. Fresh asphalt reflects approximately 4% while worn asphalt reflects about 12%. Grass reflects 25%, desert sand 40%, and concrete 55%. The term, "albedo" refers to this reflectivity percentage and is thus a dimensionless number. Based on the previous percentages, the albedo of grass is .25, desert sand is .4, concrete is .55, and so forth.

Albedo is regulated by several factors. First is the object's ability to absorb light. A black object absorbs virtually all the energy from the visible spectrum, while a white object absorbs virtually none. Most objects absorb certain wavelengths of light while reflecting others. The color of an object is actually observed as the wavelengths of light reflected rather than absorbed.

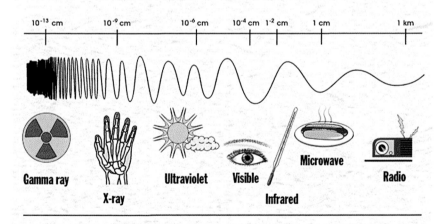

Figure 15. Diagram of the electromagnetic spectrum.

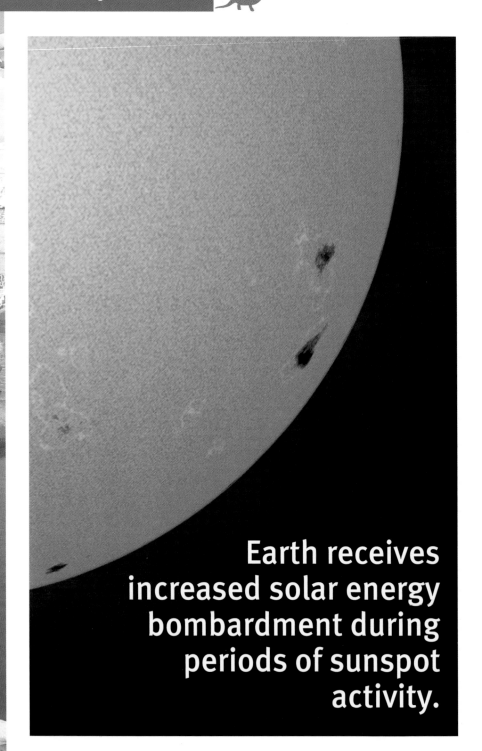

Earth receives increased solar energy bombardment during periods of sunspot activity.

Thus, a blue object absorbs all wavelengths of visible light except blue.

Visible light is only a small portion of the electromagnetic spectrum. As can be seen from figure 15, the electromagnetic spectrum includes wavelengths from gamma to radio. The shorter the wavelength, the more energy that is carried per "packet" or photon. Thus, it is much more important to reflect photons of blue through ultraviolet wavelengths than it is red through infrared wavelengths if you are trying to keep an object (such as the Earth) from heating up.

Albedo is also affected by the angle at which the light strikes an object. It is easier for an object to reflect energy impacting it at an angle than it is energy coming at it head-on. This feature is easily observable by looking at a pool of still water on a sunny day. If you look straight down, you can easily see the bottom of the pool. However, if you look at it from an angle, all you see is the reflection of the sky.

The albedo of the Earth is approximately 0.3 (Kiehl and Trenberth 1997). This means that 30% of the Sun's energy that hits the Earth is reflected back into space, while approximately 70% is converted into other forms of energy before being radiated into space. The albedo of the Earth is very difficult to measure because of the differences in its surface and its constantly changing conditions. The Earth's albedo changes from summer to winter, from varying farming practices, from visible cloud cover, from the changing amounts of forest cover, and from a multitude of other factors.

The cloud factor and other interactions allow the Earth's albedo to be self-regulating. In simplest terms, this self-regulation can be thought of as follows: the warmer the Earth, the more cloud, and the more cloud, the more reflectivity and thus the more cooling. The interaction between heating and cloud provides a balance. Atmospheric scientists will undoubtedly cringe at this greatly simplified explanation, but it serves to provide the reader with the general concept of albedo balance in Earth's atmosphere.

Albedo is the second-most important factor in Earth's temperature balance. Only the Sun's output has more influence on Earth's temperature. Albedo is also the most important factor in this balance that humans can influence. It has been suggested that farming and rainforest removal have each altered the Earth's albedo so as to contribute to global warming (Feddema, et al, 2005). The pavement and buildings of urban areas may also change the Earth's albedo by a measurable amount.

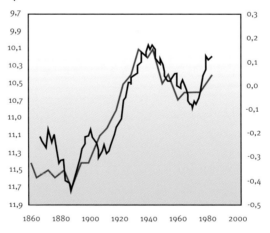

Figure 16. Variations in solar activity (red curve) and the Earth's average temperature (blackcurve) over the past 150 years.

Sunspot Activity

Most people are familiar with the 11-year cycle of sunspot activity and the fact that the Earth receives more energy during the 5½ years of increased sunspot activity than during the 5½ years of decreased sunspot activity. This cycle is actually tied to a 22-year cycle of the Sun's magnetic polar reversals. These various cycles of solar energy amplitude provide the single greatest source of energy and of temperature variation to the Earth (Trenberth and Stepaniak 2004).

The sun has a set of strong magnetic poles for only a few years during the 22 year cycle (Moussas et al 2005). During these strong magnetic periods, the Earth receives less solar radiation due to the Sun's own energy-trapping abilities. As the Sun's magnetism weakens, the Earth receives comparatively more radiation. At its weakest, magnetic poles form in many positions all over the surface of the Sun. These are what we know as sunspots. Because the Earth is occasionally positioned in line with those spots, and because that position is where solar energy escapes easiest from the Sun's own activities, the Earth receives increased solar energy bombardment during periods of increased sunspot activity.

The Sun's magnetic polar reversal cycle begins with the Sun's magnetic field at its maximum, and the Earth receiving the minimum solar energy. Over the next 5.5 years, the Sun's magnetic field gradually weakens, and the Earth receives correspondingly more energy. At 5.5 years, the Sun loses its magnatic polarity entirely, only to have the magnetic strength increas once again over the next 5.5 years to a maximum magnetic strength with the polarity reversed. This half-cycle is duplicated over the next 11 years to return the Sun to its original magnetic polarity and strength. Since solar output and sunspot presence only rely on magnetic strength and not polarity, this 22 year solar cycle produces the 11 year sunspot cycle observed from the Earth.

Evidence of the effects of sunspot activity on Earth's global climate can be seen when one compares the graphs for the number of observed sunspots versus average global temperature. As seen in Figure 16, the two graphs are virtually identical. Since sunspot activity is an indicator of the amount of energy given off by the Sun, the graph also verifies the close relationship between the amount of solar energy received by the Earth and the Earth's global average temperature.

The variation in sunspot activity during extended periods in Earth's recent history has been the focus of significant investigation. This phenomenon was first recognized by solar astronomer Edward W. Maunder (1851–1928). By reviewing the records of earlier solar astronomers, Maunder discovered the apparent lack of sunspots during the time of abnormally cool temperatures stretching from approximately 1645 A.D. to approximately 1715 A.D. Maunder noted that fewer than 50 sunspots had been recorded when there should have been hundreds or thousands. This period of time has been designated the "Maunder Minimum" in his honor.

Other researchers have identified several other periods when sunspots were rare, and each correlates with a global cooling event on Earth. The reason the Sun exhibits periods when sunspots are plentiful and other times when they are virtually non-existent is unclear. However, these phenomena appear to be related to the fact that different portions of the Sun rotate at different speeds. Whatever the cause, the number of sunspots is an accurate gauge for the amount of solar energy being given off, and thus also an accurate gauge for energy reaching the Earth.

Milankovitch Cycles

The amount of solar energy reaching the Earth is not only regulated by the amount of energy being given off by the Sun, but also by the Earth's positioning relative to the Sun. It was only during the past 100 years that scientists have become aware of how important this factor is in reference to the Earth's ability to capture solar energy.

Serbian civil engineer and mathematician Milutin Milanković revolutionized our understanding of the effects of the eccentricity, axial tilt, and precession of the Earth's orbit on global climate (Boggs 2001). It is beyond the scope of this book to explain the details of these interactions, but a summary is justified. It seems that when the Earth's orbital "spin and wobble" characteristics are plotted against global temperature averages, the combinations of these factors most likely to allow the maximum solar energy to reach the Earth correlate strongly with observed global temperature increases.

The Earth's eccentricity, or deviation from a perfectly circular orbit around the Sun, changes through time as a result of gravitational interactions, primarily with Jupiter and Saturn. If the only gravitational effect on the Earth was that of the Sun, the eccentricity would not vary. However, due to the changes in gravitational fields depending on orbital alignments of these other planets, the Earth's eccentricity varies through periods of lower and higher eccentricity. Thus, the Earth goes from being an average distance from the sun throughout its orbit (low eccentricity) to being closer and farther away (higher eccentricity) and back again throughout this cyclic period. This forms a pattern that repeats roughly every 100,000 years or so.

The Earth's axial tilt also varies. The Earth's axial tilt forms an angle of approximately 21½ degrees from perpendicular at its minimum to 24½ degrees at its maximum. When the angle is at its maximum, the differences between summer and winter are the largest. However, when the angle is the smallest, there is less impetus for a change in seasons. For this reason, some scientists argue that it is easier for polar ice caps to form when the axial tilt is minimized (Berger 1988). The pattern of axial tilt variation repeats every 41,000 years.

The third variable in Earth's orbit around the Sun is the Earth's precession, or "wobble". The Earth's axis circles around its perpendicular just like a spinning top's axis circles around its perpendicular as it slows. The effect of this "wobble" is that, over a 23,000 year period, the Earth goes from being closest to the Sun when it is summer in the northern hemisphere, to being farthest away at that time, and back to being closest again. The effect that this has on Earth's climate can best be explained by example. For instance, it would be much more difficult to maintain a glacier on the Antarctic continent when the southern hemisphere is closer to the Sun in the winter. Currently, the combination of the Earth's axial tilt and its precession make the Earth farthest away from the Sun during Antarctic winter (Berger et al 2003). Over the next 5,500 years, the northern and southern hemispheres will become equidistant from the Sun during their winter periods.

These three variations with three different periods interact with each other to amplify or nullify each other's effects. Although these variations seem small, they change the total amount of sunlight reaching the Earth by up to 25% at mid-latitudes. As such, they play a huge role in Earth's global climate. The good news for those of us who would hate to see glaciers covering Europe, Canada, and Russia is that most scientists who study the Milankovitch cycles believe that the current interglacial period will last for another 50,000 years before the Earth starts cooling once again.

Human Population, Activity, and Growth

While this topic is neither one of the factors contributing directly to global climate regulation nor a focus of this book, any list of factors affecting Earth's temperatures at present and in the future would be incomplete should human population dynamics not be considered.

Within the next 30 years, the Earth's population may increase by 50% (Cohen 2003), and standards of living are increasing. The use of powered, individual transportation and electricity are increasing more rapidly than the human population itself. Even though major breakthroughs in efficiency and production have occurred, unless conditions change drastically, a great deal more energy PER PERSON will be needed to supply the new living standards. Added to the increases in population, this extra energy cannot be provided within current human activity patterns without increasing the use of fossil fuels.

One of the issues that is constantly in the news is the problem of burning fossil fuels for transportation. Figure 17 shows that, while transportation is a significant contributor, its contribution is only a small portion of the total picture.

We are already at the limit where gasoline production and use are balanced – we are not finding enough new reserves to provide for the demand, and what we do find cannot be refined quickly enough. While ethanol has been added at up to 10% in automotive fuels, this has only resulted in about 5% of the U.S. domestic fuel being from renewable sources. The next big "push" is in the form of electric vehicles, where transportation will gain its energy from the electric grid rather than from gasoline. However, it should be recognized that, according to the U.S. Energy Information Administration, 2/3 of the electricity in the U.S. grid comes from burning coal, natural gas, and diesel fuel. As the electrical demand increases, coal-fired electrical generation, as the only major 24/7-available source, will bear the brunt of the increase.

According to the most recent data on carbon emissions shown below, transportation is shown to be nearly equal to electrical generation in terms of emissions. However, these numbers are skewed. What is shown is the emissions after the exhaust from electrical generation is "scrubbed." Scrubbing is done a number of ways, but the bottom line is there is additional carbon product that needs to be sequestered. In several of these processes, the disposal allows carbon to recirculate in the biosphere, and one – biologic scrubbing, produces organic material that is then fed to livestock. While these processes keep CO_2 from being directly added to the atmosphere, over time the coal-carbon atoms still end up as components of methane and CO_2. Unfortunately, the government reports consider only the primary disposal, and not the entire cycle to sequestration.

Yet these sources are under-utilized. For example, Fort Peck Dam in Montana only has the ability to generate electricity from 10% of its peak flow, and it typically generates at only a small portion of its current capability. Were its total generating capacity, including full flow, to be utilized, the electricity could replace that currently being produced by a large coal-fired plant. Currently operating hydroelectric plants are capable of supplying ap-

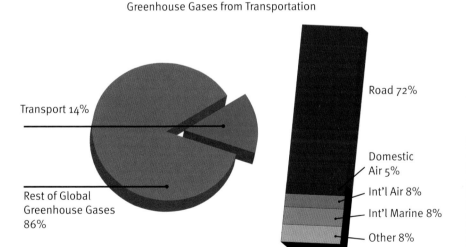

Greenhouse Gases from Transportation

Transport 14%

Rest of Global Greenhouse Gases 86%

Road 72%

Domestic Air 5%

Int'l Air 8%

Int'l Marine 8%

Other 8%

Figure 17. Global total carbon use, from transportation.

U.S. Energy Consumption (Quadrillion Btu)
Note: Expressed as gross calorific values.

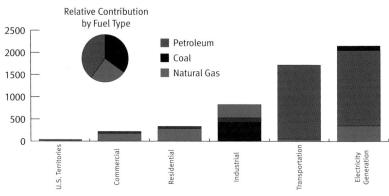

Figure 18. U.S. Energy consumption.

proximately 14% of demand, but are in reality only supplying less than half that amount. In addition, geothermal electrical production – the process of pumping water into deep wells where it heats into steam, then using the steam to drive turbines – is virtually an untapped resource.

It is important to understand that all of the sectors listed in Figure 18 use previously sequestered carbon and add it to the atmosphere/biosphere. Technology does exist to provide energy that doesn't rely on these fossil resources. Yet these sources are underutilized. For example, it is a sad commentary that, in what has long been considered the most progressive and affluent country in the world, the two dependable, 24-hour, clean sources of electricity are among the smallest percentage provided to the grid. Hydroelectricity has virtually no downside. Neither does geothermal technology.

Both are capable of providing power to the grid at all times, unlike solar or wind power.

Providing energy for heat, lights, manufacturing, food production, transport, and a host of other activities to supply a doubled population will strain our infrastructure to its limit. China faced this problem right after World War II and has endured much criticism for its stance on mandatory limits on childbirth. However, without such limits they could not have prevented massive starvation and societal collapse within their country. With limited resources, a nation can only feed a certain number of hungry mouths. China may have taken extreme and potentially unjust measures to achieve its goals, but the end result is that their population currently has enough resources to survive and thrive. If they had done nothing, starvation and disease would have killed millions and jeopardized the nation itself.

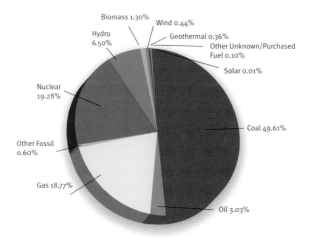

Figure 19. Sources of energy supplied to the U.S. electrical grid and their relative percentages.

The world itself has limited resources. It can barely provide food and energy demands at current levels to our existing population; it will be much less able to support a doubled population with greater energy needs in 30 years. Of great concern will be the initial scramble for the remaining natural resources and especially the fossil fuels. Once those resources are gone, what will replace them? And if we throw that much carbon into the atmosphere, what will be the effect on the planet's energy balance and climate?

Unfortunately, humanity's need to provide energy to its individuals is directly opposed to its need to prevent climate catastrophe. Only through foresight, careful planning, and drastic cuts in both population increase and fossil fuel use can humanity hope to survive past the next 3 generations and still maintain the abilities for healthy food, proper heating/cooling, and travel needs. Understanding Earth's energy balance, and what upsetting that balance could cause, needs to be a major portion of the foundation upon which humans plan for their future survival, society, and infrastructure.

The world itself has limited resources.

Chapter Four

Earth's History:
A Short Course

A scene from the Jurassic

Many books have been written on Earth's past, and the wealth of data used to reconstruct Earth's history is staggering. It would be impossible to summarize this data in detail in this volume. The following are highlights and generalizations. However, these particular topics provide the key to understanding our current climate issues and, perhaps, also provide the key to preventing the imminent extinction of the human species.

The geologic dates provided in this book are from the 2004 stratigraphic report produced by the International Commission on Stratigraphy – the worldwide organization dedicated to providing the most current information on age dates for the Earth's rock layers (Gradstein, F.M. and J.G. Ogg, et al., 2004).

According to the most recent data, the Earth coalesced from a dust cloud more than 4.55 billion years ago. The Earth has been age-dated by over 50 different, independently derived methods. Originally, ages were inferred from estimating the time it took to build up sediments that now form the rocks of our continental crust. Next, scientists used the fossil content of those rocks and species turnover as an estimate of time. More recently, a number of directly quantifiable methods have been used. These methods are based on the radioactive decay of uranium to lead, potassium to argon, rubidium to strontium, and samarium to neodymium, as well as counting fission tracks in quartz crystals. In addition, advanced techniques such as measuring iron-rich rocks and calculating the number of polar reversals the Earth has undergone since the rock was formed provide data for shorter periods of geologic time – say the past 100 million years or so. Lastly, astronomers have used various techniques to calculate the age of the universe since the "big bang" and have provided significant independent corroboration of age dates for the origin of the Earth and the Moon.

Earth's Hot Origin and Early History

According to current interpretations (Bouvier et al. 2007; Wada and Kokubo 2006), just as the Earth was becoming stable somewhere around 4.5 billion years ago, another planet roughly the size of Mars collided with the Earth. This collision did a number of things that left permanent effects behind. First, it knocked the Earth's axis off its original center, and after its rotation and revolution stabilized, the Earth's axis was tilted by an approximate average of 23.5 degrees.

Second, it blasted away a significant portion of the material and launched it into orbit, providing the raw molten mass that eventually coalesced to form our moon. Third, it imparted a "spin" to the planet that was originally about 5 hours per revolution. The Earth's rotation has been slowing down since then at a regular rate to where it now takes approximately 24 hours per rotation. And finally, the solid center of the other proto-planet sank to the center of our planet, combining with Earth's own central core and heating the Earth tremendously. This heat energy released from the impact resulted in the Earth above the central core to again rise to molten temperatures.

For the next half-billion years, the Earth slowly cooled. Perhaps as early as 4.3 billion years ago, but certainly by 4.05 billion years ago, Earth's surface had cooled sufficiently for continents and oceans to form (Wilde et al 2001). At that point, Earth's atmosphere was composed primarily of ammonia, methane, water vapor, carbon dioxide, and nitrogen. The Sun's output at that time was only 60% of its current energy output, and this factor allowed the Earth to cool more rapidly than it would today.

The temperature balance of the Earth, then as now, was and is a balance between heat rising to the surface from the Earth's interior and energy striking the Earth from space being offset by energy being radiated from the Earth into space. Primitively, Earth's temperature balance included much more energy radiating from Earth's interior and much less being absorbed from the Sun. Since a faster radiative cooling can occur if the energy is only

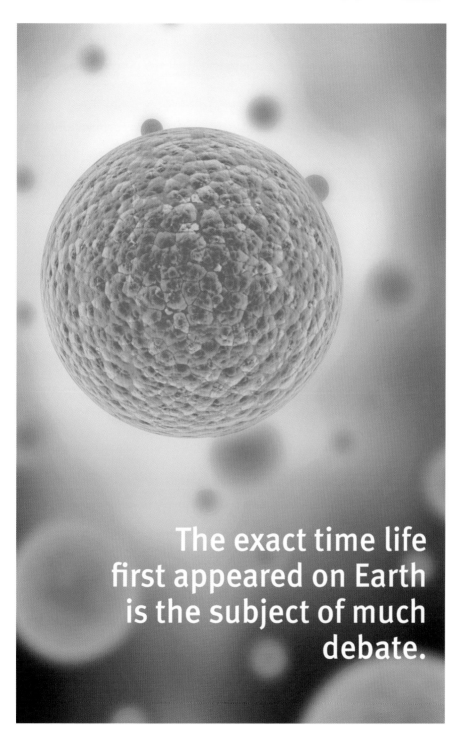

The exact time life first appeared on Earth is the subject of much debate.

being transported outward rather than being absorbed, transformed, and re-radiated, the early history of the Earth provided a better potential for cooling than would be expected to occur today.

Another factor responsible for an increased cooling rate is the formation of liquid water. Water boiled on Earth's surface gathers heat energy, transports it upward into clouds, and radiates it into space as the gas condenses back into a liquid state. This process is why it works when we pour water on an object to cool it. It worked back then on a global scale as well.

This early part of Earth's history is that of an initial, geologically rapid, cooling followed by the stabilization (within fairly broad parameters) of a climate regulated in a large part by Earth's oceans. Since the initial stabilization, the Earth has maintained average ocean surface temperatures between 70 degrees Celsius (158 degrees Fahrenheit) and 4 degrees Celsius (39 degrees Fahrenheit). Bear in mind that these temperatures do not reflect the average atmospheric temperature, since the temperature estimates come from the upper portion of the ocean. Neither do we currently identify data that would indicate the Earth's surface temperatures until plants became commonplace on Earth's terrestrial surface a few hundred million years ago.

The exact time life first appeared on Earth is the subject of much debate. If one uses purportedly life-formed carbon molecules (that have also been found in meteorites) as evidence of life, then life on Earth is as old as the oldest sedimentary rocks at 4 billion years before present. However, some researchers believe these molecules may be formed by inorganic processes, so the supposition that life existed, based on only these molecules, is tenuous (Wacey et. al. 2009).

A more suggestive bit of evidence of life is that of banded iron formations. Such banding is thought to have been produced by intermittent increases in free oxygen in the biosphere, and this free oxygen came from living organisms. Banded iron formations appear in the fossil record approximately 3.8 billion years ago, or 200 million years later than the earliest carbon molecule evidence (Fortin and Langley 2005). Fossils themselves – the direct imprint of obvious once-living structures - have been found in sediments 3.5 billion years old (Tice and Lowe 2004). These fossils are the preserved remnants of mats of algae called stromatolites. Evidence of complex life – animals with internal organs – appeared in the fossil record a mere 600 million years ago.

One question that has been largely unexplored is, "Why does complex life show up so late in the geologic record?" Discoveries made in just the past 10 years or so may shed some light on this question. The most likely answer, in a nutshell, is that complex life likely could not have survived on Earth for much of its history. The Earth's climate and atmospheric conditions were too inhospitable.

For nearly the first half of Earth's history, its atmosphere contained little or no free oxygen (Kasting and Howard 2006). Most complex life forms, other than photosynthesizing organisms, depend on free oxygen to drive their organic reactions. Thus by definition, complex life, at least as we know it, could not have existed prior to the availability of free atmospheric oxygen in significant concentration. Also, ozone is a by-product of free atmospheric oxygen, so the Earth could not have had an ozone layer before there was a significant percentage of free oxygen in the atmosphere. Early life would have thus been subjected to much higher levels of radiation than we are at present, again making it difficult for complex life to survive at that time.

Archaea – organisms that, in most cases, need neither sunlight nor oxygen to survive - can be found living in a wide variety of environments. Lack of sunlight, lack of oxygen, extremely high or low temperatures, and even extremely salty or chemically rich environments are not barriers for survival of species within this group. Their metabolic processes vary so widely that consistency between various groups of these organisms is nonexistent. Our knowledge of these organisms comes only from the past few dozen years, and even for most of that period scientists relegated this group to extreme environments. We now know, however, that these organisms are found in virtually all environments on Earth. It is thought that members of this group were among the earliest life forms on Earth.

It is interesting to note that the most likely source of free oxygen in our atmosphere is as a waste product of the metabolism of carbon dioxide. It was initially produced by photosynthesizing cyanobacteria and later by

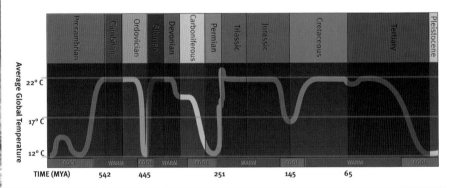

Figure 20. Global temperature curve for the past 600 million years.

more complex photosynthesizing organisms, though the actual increase in atmospheric oxygen concentration may be a more complicated story (Kump, 2008). Once organisms that could use solar energy to drive metabolic reactions developed a few billion years ago, a whole new process began. This process allowed those organisms to "eat" carbon dioxide – a waste product of the earlier life forms – and release free oxygen. In turn, once free oxygen existed in quantity in the atmosphere, organisms developed that could use oxygen to drive their metabolic reactions, once again giving off carbon dioxide as a waste product.

Fortunately for us, an equilibrium has been established whereby plants that release oxygen into the atmosphere balance animals that use it, and through this balance both these forms of life survive and thrive. However, it has only been in the last few hundred million years where this balance has been established, and previous imbalances were likely involved in earlier catastrophic environmental and ecosystem collapses.

Our current biologic balance on Earth consists mostly of organisms that regulate their quantities and activities, or rely on other organisms to regulate their quantities and activities, such that a balanced carrying capacity is maintained. Organisms such as amoebas who reproduce indiscriminately do exist, and they increase in number until a critical resource in their environment is used up. This causes entire populations to die off, to be replaced only by migrating individuals from other areas/habitats once the previous habitat is re-established. But these are the exception rather than the rule. Through time, organisms that poisoned their environments or overpopulated have been virtually eliminated from the biosphere in favor of those that can maintain their environments over the long term.

It is central to our understanding of current processes and problems that virtually all free oxygen existing in our atmosphere today is the result of photosynthesis. The original match of free oxygen to free carbon and the subsequent recombination of oxygen with iron and other elements means that there exist many more free carbon atoms sequestered somewhere within the Earth's system than there are free oxygen molecules. Therefore, we will run out of atmospheric oxygen long before we burn all of Earth's sequestered carbon resources. Already, we have diminished atmospheric oxygen amounts measurably through our burning of fossil fuels.

The Earth Blows Hot and Cold, and Severe Extinction Events Occur

About 2.9 billion years ago, the Earth began experiencing its first ice age. This event is coincident with the first major upheaval in the Earth's atmospheric makeup (Kasting and Howard 2006). Cyanobacteria and other photosynthesizing organisms were removing carbon dioxide from the atmosphere at a tremendous rate, and this was likely a factor in the global cooling. This first ice age lasted approximately 400 million years and ended at virtually the same time free oxygen showed up in Earth's atmosphere. Little is known about this ice age other than it was severe and long-lasting.

Not all ice ages are equal in severity or duration. For the purposes of this discussion, the term, "ice age," is defined as any time when large polar ice caps and significant temperate latitude glaciation exists. Thus, by definition, we are currently in an ice age.

Since the ice age of 2.9 billion years ago, the Earth has experienced four additional major ice ages. The most severe was the next in line at approximately 850 million to 544 million years before present. This event, from evidence gathered so far, seems to have seen virtually the entire Earth covered with ice intermittently throughout that 220 million year period (Rieu et al 2007). If it hadn't been for carbon dioxide thrown into the atmosphere by volcanic eruptions and non-photosynthesizing organisms such as extremophile bacteria producing carbon dioxide as metabolic waste, there would have been no mechanism to cause global warming, and the Earth would have remained covered in ice to this day.

It is at the end of this glaciation event that complex life appears in the fossil record. It is also at the end of this glaciation event that the first rapid temperature spike – where global temperatures rose instantaneously (from a geologic perspective – defined as "below observable resolution in the rock and fossil records") – occurred. Recent evidence links this temperature spike with the rapid global melting of methane clathrate. The production of this substance and its impact on the current and future biosphere will be discussed in more detail in the next chapter.

Complex life leaves more in the rock and fossil record than simple forms.

Once complex life appeared on Earth, the documentation of extinction events becomes much easier. Complex life, by definition, leaves more and better indications of their presence in the rock and fossil records than simpler forms. Hard body parts contribute tremendously to our ability to document the presence or absence of these organisms at any given time in Earth's history. In turn, it is the perceived patterns of their presence/absence that we rely on to provide the information necessary to create a plan to prevent our own extinction event. Throughout the rest of this chapter, we will examine significant global extinction events and concurrent global climate changes, since the two are demonstrably correlated.

The first significant glaciation event after complex life appeared occurred from 460 to 430 million years ago. This relatively short-lived glaciation is coincident with the then-continent of Gondwana passing over the South Pole (Sheehan 2001). Glaciers formed over much of the Gondwanan landmass during that period. The formation of these glaciers caused global sea levels to drop significantly and then fluctuate as the ice formed and melted repeatedly. This event caused more loss of species than any other extinction event save that at the Permian/Triassic boundary 251 million years ago.

The severity of this extinction is likely tied to its effect on species that couldn't move quickly enough to a habitable environment. Corals and other complex organisms attached to the ocean floor were unable to migrate, so when the ocean became too shallow or too deep, they died. It is important to note that virtually all life on Earth at this time lived in the marine environment, and many of the more mobile forms relied on the sessile organisms for their nutrition and habitat. It may be that this particular event would not have been particularly damaging to Earth's biota had the majority of complex organisms been, in general, more mobile. It is possible that evolutionary adaptations to overcome this factor allowed organisms of the later Paleozoic to migrate into terrestrial habitats, encouraged walking, running, and flight, and perhaps even promoted incipient warm-bloodedness.

A series of minor glaciation events beginning approximately 350 million years ago culminated with a larger glaciation event in the late Permian, approximately 253 million years before present. Except for the more significant glaciation at the end of the Permian, these events were of only a few million years' duration and were separated by tens of millions of years. Not only were they of short duration, but they were also mostly restricted to polar and sub-polar regions. During most of the later Paleozoic, subtropical climates were the norm throughout the middle latitudes. The Permian is notable in that, like now, the Earth had experienced a gradual cooling cycle followed by a gradual warming until temperatures suddenly spiked (e.g. Kiehl and Shields 2005; Ryberg and Taylor 2007). During this time, glaciers often covered one or both of the Polar Regions at the same time deserts and tropical climates were found near the equator.

Species numbers were in gradual decline throughout the latter portion of the Permian, just as they have been in decline the past few million years of our recent history. However, evidence exists that most species were lost during a period of not more than a few tens of years coinciding with a temperature spike at the Permian/Triassic boundary 251 million years ago (Dorritie

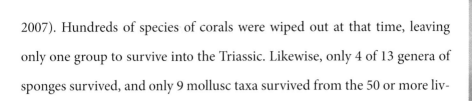

2007). Hundreds of species of corals were wiped out at that time, leaving only one group to survive into the Triassic. Likewise, only 4 of 13 genera of sponges survived, and only 9 mollusc taxa survived from the 50 or more living in the Permian.

Interestingly, terrestrial plants and vertebrates suffered least from this extinction event. The fact that the supercontinent, Pangaea, existed at that time was likely a factor in terrestrial species survival. Isolated remnants of populations could quickly regroup, migrate, and repopulate regions as they recovered from the devastation. Also, there were no barriers to prevent animals from migrating to the far north or far south so as to avoid the persistently intense heat that would have been present in the tropics.

The Permo-Triassic event is followed by millions of years where the global temperatures of the Earth were significantly warmer than average (Retallack 2005). Year-around ice caps at the Earth's poles were not present for at least 100 million years following this event, and large scale polar ice caps did not exist for over 200 million years. Temperatures on the Earth after the Permian/Triassic event did not again lower to Early Permian averages until the Miocene of approximately 20 million years ago - 230 million years later.

Dinosaurs roamed the Earth for over 150 million years, beginning in the early Triassic. Contrary to popular belief, however, individual dinosaur species did not endure for more than a few million years within that time. The history of the dinosaurs is one of species evolving and dying out, only to be replaced by other dinosaur species. At first, dinosaurs were generalized creatures quite similar to their bipedal crocodilian relatives. Crocodiles, like snakes and whales, can trace their ancestry to relatively long-legged, fast-moving forms. Some early crocodilians were the first vertebrate life forms to walk upright on two legs, and it was from this lineage that dinosaurs and birds arose. Dinosaurs survived two major extinction events in their sojourn on Earth, but they succumbed to the third major extinction event they faced. Known as the K/T (shorthand for Cretaceous/Tertiary) boundary, this event has caused more speculation than any other extinction event.

Contrary to popular belief, dinosaurs weren't doing fine until a rock from space wiped them out. While there is no doubt that the asteroid that struck Mexico's Yucatan Peninsula caused significant climate disruptions, there is no direct evidence that this event killed any dinosaurs. Geologically simultaneous occurrence does not prove cause and effect. And there is some evidence to suggest virtually all dinosaurs died out a half-million years before the event. There is also evidence to suggest some dinosaurs may have survived the event. Dinosaurs most likely fell victim to their own specialization and reliance on a stable environment, and they could not adapt when that environment changed.

10 million years before the demise of the dinosaurs, an estimated 400 different dinosaur species roamed the Earth. By 1/2 million years before their end, the number of species had dwindled by at least 75% (Weishampel et al. 1990). The animal population numbers, however, had remained constant or even increased. What this tells us is that certain species of animals got very good at doing what they did to survive – so much so that less-successful forms could not compete and thus died out. The animals best suited to that

The history of the dinosaurs is one of species evolving and dying out, only to be replaced by other dinosaur species.

stable environment increased in numbers to the point that there was no food or room left for the less successful forms.

Such specialization works well until the environment changes. At the end of the Cretaceous, the dinosaurs faced a climate change that lowered the average temperatures and allowed a new, and likely indigestible, species of plant – grass - to become the dominant ground cover. The dinosaurs were also subjected to a series of catastrophic events including massive volcanic eruptions and several large meteor impacts (e.g. Zhao et al. 2009). They even experienced snowstorms and cold - the likes of which dinosaurs had never before faced.

With the more generalist species of dinosaurs out-competed earlier, this left only the specialized forms to deal with these rapidly changing conditions. It is likely that even the specialized dinosaurs could have survived any one of these events, but they could not adapt to the rapid changes caused by all these events occurring simultaneously.

Thanks to that same series of events, however, mammals – who had for 100 million years lived in the shadow of the more efficient dinosaurs – now had a chance to become masters of the Earth in their own right. Huge titanotheres, giant sloths, saber-toothed cats, and precursors to camels, horses, bison, elephants, and bears roamed the prairies just a few million years after the dinosaurs became extinct. Mammals quickly became diverse and plentiful once the End-Cretaceous disasters destroyed the dinosaurs.

Approximately 10 million years after the dinosaurs' demise, the Earth once again experienced a relatively sudden temperature spike. Sea surface

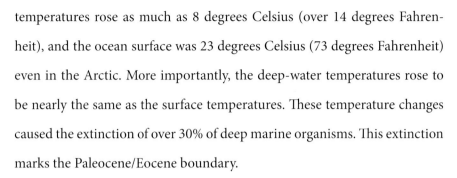

temperatures rose as much as 8 degrees Celsius (over 14 degrees Fahrenheit), and the ocean surface was 23 degrees Celsius (73 degrees Fahrenheit) even in the Arctic. More importantly, the deep-water temperatures rose to be nearly the same as the surface temperatures. These temperature changes caused the extinction of over 30% of deep marine organisms. This extinction marks the Paleocene/Eocene boundary.

This catastrophic rise in Earth's temperature has been linked to methane clathrate breakdown (Katz et al 2001). An increase in atmospheric carbon is noted from this time through several separate lines of evidence. This event marks the last significant release of the methane trapped on the ocean floor. Temperatures on Earth gradually cooled from the Paleocene/Eocene Thermal Maximum (PETM), but it took over 10 million years for temperatures to decline to where they were prior to the event.

Approximately 40 million years after the dinosaurs' demise, Earth's climate began cooling in earnest. Once again, a changing climate caused the extinction of the more specialized species and forced adaptation allowed the survival of the more generalized forms. This cooling ultimately led to the formation of polar ice caps and cooler, drier conditions over most of the Earth.

The last ice age began in the Miocene epoch, approximately 23 million years ago, and has lasted to this day. The major glaciation events of this last ice age, however, have occurred within the last million years. The Earth today is still significantly colder than its geologic average, and even the early Middle Ages experienced warmer climates than we are witnessing at present.

The Earth is, at present, roughly 10 degrees Fahrenheit (6 degrees Celsius) cooler than its geologic average (Lisiecki and Raymo 2005). Were the Earth at its equilibrium point, based on geologic and paleontologic data, the polar ice caps would be melted and the middle latitudes worldwide would be subtropical!

Recent studies, such as that of Lisiecki and Raymo (2005), indicate that over the past 5 million years, the Earth has become colder than it has been at any time in the past 250 million years. We are in the middle of an ice age the likes of which the Earth hasn't experienced since the Permian Period. Even the sudden hikes in global temperatures of the past few years have not raised temperatures globally back to the point they were when the Tower of London was built and when the First Crusade took place. To overcome the unnaturally low global temperatures and bring things back to Earth-normal, the global temperature average must raise another 10 degrees Fahrenheit.

From the diagram in Figure 21, it is plain to see that the temperature fluctuations have become much greater over the past 3 million years than they were previously. Also, the average ocean temperature has dropped approximately 5 degrees Celsius (9 degrees Fahrenheit) during the same period. However, it can also be seen that the most recent data points place the current temperatures at or near the earlier average temperatures before the cold incursion.

As shown in Figure 21, the Earth's temperature fluctuates around a mean temperature, and the amplitude of this fluctuation has been increasing. The warmer the high in relation to this mean, the lower the corresponding low. It

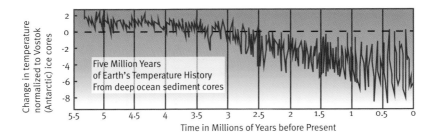

Figure 21. Changes in ocean temperatures over the last 5 million years, as indicated by stable isotope data.

is this cyclic nature that has led some scientists to predict that the current warm spell is going to trigger the next ice age. This scenario would be likely except for the increase in greenhouse gases not present in the previous cycles.

In all, it appears that the Earth's current temperature rise marks a return to a moderate temperature range from an exceptional cold period as has happened many times in the past. It now appears unlikely that the cold temperatures most of us still remember from our childhood will return anytime soon, and we may yet in our lifetimes see crops grown in Greenland such as were in the days of Leif Eriksson and Eric the Red of Norse legend. Imagine – we are upset that the climate is now the same as it was a thousand years ago!

This Violent Planet

As a matter of perspective, the Earth, through time, is an unstable, changeable place. Its climate and ecosystems are periodically readjusted to a balance that seems to maintain subtropical environments throughout the middle latitudes, and where polar ice caps cannot exist. The Earth tends to cool slowly, then warm rapidly and violently. Each warming event is accompanied by a concurrent massive extinction. In between, the Earth's temperature balance may fluctuate globally a few degrees such as it has been doing over the past million years or so. These minor fluctuations have produced the "recent" glacial and interglacial periods.

The last major readjustment from cold to warm occurred 251 million years ago, and during that process over 90% of all life was wiped off the face of the Earth. Conditions today mirror those found then – just before the catastrophic extinction. The balance of this book is dedicated to the examination of that event, its causes and effects, and to what we can and cannot do to prevent a similar catastrophic event from happening in the near future.

The Earth, through time, is an unstable, changeable place.

Chapter Five

The Great Dying

Parched Earth

Over the last 650 million years, there have been three major global warming events and one minor one. These events occurred approximately 650, 450, 250, and 50 million years ago. The first three corrected millions of years of ice ages and returned virtually the entire Earth to a tropical environmental regime. The last also raised the Earth's temperatures into the tropical range, but the Earth was relatively warm to start. The three major global warming events listed above coincide with three of the largest extinction events in Earth's history as defined by the percentage of species lost during the event.

Admittedly, the data needed to understand all these events in detail is just beginning to be gathered. The data we have been able to collect, however, provide us with a cyclic pattern that is pertinent to our current existence. More importantly, the data are better, and the picture clearer, for the events of 260 to 240 million years ago, and these data show striking parallels between the Earth 251 million years ago and the Earth today.

The greatest loss of life the Earth has ever experienced occurred at the end of the Permian Period, and by its culmination over 90% of all life on Earth was wiped out. This event has been referred to as "The Great Dying." And it is very clear from all available data that this extinction was driven by rapid climate change.

During the end of the Permian Period, the Earth experienced a gradual cooling, intermittent ice ages with slightly warmer interglacial periods similar to those of the past 2 million years, and then a gradual warming such as Earth has been experiencing over the past few thousand years. Then, and unlike our recent climate history thus far, the Earth experienced a spike where temperatures warmed catastrophically (Rees 2002; Retallack 2005; Shen et al. 2006). There had been an ongoing die-off of species during the last half of the Permian Period just as we are seeing today, likely due to similar geologically rapid variations in climate and changes in sea levels. However, geographically the Earth was very different from present - there was one supercontinent, Pangaea, in the late Paleozoic. Ocean circulation patterns were also entirely different.

In terms of global climate and vegetation, the late Permian Period was broadly analogous to that of today. There were polar icecaps and equatorial rainforests, albeit populated by very different kinds of plants. There were large desert areas in the low- and mid- latitudes, possibly corresponding to our current boundary area where Hadley Circulation meets up with mid latitude circulation patterns. The single landmass would have created much simpler, and thus much broader, patterns for ocean circulations, air flow, and energy transport patterns, and these may have contributed to the larger Late Permian desert areas. It is well documented that the energy/temperature gradient between tropics and Polar Regions was significantly less then as well (Ryberg and Taylor 2007; Rees 2002). In other words, the tropics were cooler and the poles warmer, relative to each other, than the equivalent regions today.

As the Earth slowly warmed during the last million or so years of the Late Permian Period, approximately 252 million years ago, tropical areas became warmer and drier. Deserts formed where lush rainforests previously existed. Ice caps melted, and sea levels fluctuated dramatically between what were likely glacial and interglacial periods. Carbon dioxide levels gradually increased to approximately twice current levels, then spiked to over 4 times current levels (Retallack 2005). While there could be several explanations for this CO_2 spike, the most likely is volcanic venting related to the Siberian Traps Volcanics that covered a large portion of the Russian land area.

These same processes can be seen in today's Earth, where the African Sahara rainforest of a few thousand years ago has now become the Earth's largest non-polar desert (Antarctica is classed as the Earth's largest desert). Earth's ice caps are melting, and it is estimated that carbon dioxide levels will reach twice pre-industrialization levels as early as the year 2050

The greatest loss of life the Earth has ever experienced occurred at the end of the Permian Period.

(Intergovernmental Panel on Climate Change. Working Group I, 2001). Even if this estimate is overly pessimistic, it is still likely that CO_2 levels will reach those levels in what is left in most of our lifetimes (Glasby 2006).

These predictions are based on current activities and trends. A major infusion of carbon dioxide into the system through volcanic venting, or any of several other potential disruptive factors could greatly accelerate the atmospheric CO_2 increase rate. In any case, by the end of the century, global temperatures, polar ice cap conditions, atmospheric CO_2, and many other factors are predicted to mirror those found at the exact time of the Permian/Triassic extinction catastrophe point.

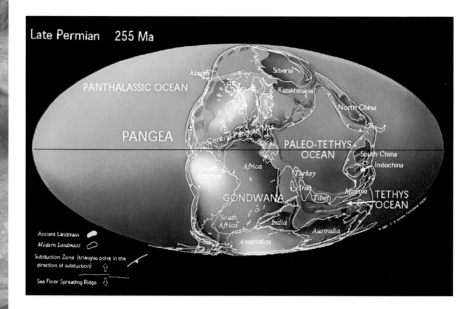

Figure 22. The continents as arranged at the end of the Permian Period.

So, what were the causes, effects, and environmental conditions across the P/T (Permian/Triassic) extinction event? Initially, the climate conditions were the same as they are at present, sans humans and including the Siberian Traps volcanics. The carbon dioxide and sulfur dioxide vented from that volcanic series put similar amounts of carbon dioxide and greater amounts of sulfur dioxide into the atmosphere compared to human activities today. The carbon dioxide caused gradual global warming. The sulfur dioxide (SO_2), on the other hand, combined with water to form sulfuric acid (H_2SO_4). This caused rapid erosion of carbonate rocks on land and an increase in the acidity of the oceans.

Atmospheric CO_2, carbonic acid, volcanic compounds, and new land-based chemicals leeched into the water system rapidly overwhelmed the oceans' chemical balance. A series of die-offs occurred, and ultimately, the oceans were virtually sterile. Coral reefs, prior to the extinction, were built by tabulate and rugose corals and by sponges. None of these reef-building animals survived the P/T event (Knoll et al 2007). Modern-type corals appeared in the mid-Triassic, at least 12 million years after the P/T event, and these animals are more closely related to sea anemones than to the reef-building animals of the Permian.

Many forms of life died out completely at the P/T boundary. However, many species were in decline for much of the Permian period. This decline was likely due to the fluctuation of sea levels from the Permian's glacial and interglacial periods. These groups include trilobites, sea scorpions, blastoid echinoderms, and acanthodian fish. Other groups seemed to be doing well up until the end, and of these, often a species or two would survive. Such are represented by radiolarians, foraminifera, corals, snails, and bivalves.

Lack of land-based sediment preservation and conditions unfavorable to fossil preservation, such as the increased acidity described earlier, has limited our knowledge of land-based events. We do know that approximately half the known animal genera disappear at the boundary, and the survival versus extinction of the land genera seems almost random. Such random survival may be linked directly to variations in intensity and/or duration of harsh environmental conditions, and terrestrial populations may have survived in certain areas simply because conditions weren't as severe as they were in other areas. The terrestrial survival patterns could thus best be described as "survival of the lucky" rather than "survival of the fittest."

Perhaps the most informative loss of land taxa occurred in the plant realm. Coal forms primarily in swampy, often tree-covered terrain. The actual carbon comes largely from moss and leaves being buried and then baked by heat and pressure. Coal deposits are exceedingly common in Pennsylvanian Period sediments. In fact, the European "Carboniferous Period" was named based on the abundant coal deposits from that time period. The formation of coal continued throughout the Permian Period in virtually all land areas that were not arid deserts. However, no coal is found in any Early Triassic sediments (Retallack et al 1996). The remains of any plants during that time are extremely rare. Part of this can be explained by the presence of large areas of desert at that time; but even in well-watered areas, coal is not found. This suggests that areas where trees and other macro plants survived were small in area and remote in location, and that it took tens of millions of years for remnants of this once-prevalent group to re-establish themselves from the isolated relict floras.

Since plants are a significant link in the food chain, the loss of such at the P/T boundary is likely a significant factor in the loss of other land organisms – especially those who relied on such for their nourishment. As we have learned from studying the Sahara, once plants are stripped from the Earth's surface, deserts form easily and recovery from desert conditions is very difficult (Saier 2010; Sivakumar 2007) . These factors combine to paint a picture, for the first few million years after the P/T boundary event, of most terrestrial landmasses being largely covered by deserts. In those areas not covered by desert, the land would still have been mostly bare of vegetation. Land animals that survived likely did so by eating most anything available. It was a time where generalists – animals that can survive in a broad range of conditions – initially have a distinct advantage. As the climate stabilized, however, specialized taxa – those who are better at surviving in a particular set of environmental conditions - were again encouraged.

By mid-Triassic, some 20 million years after the catastrophic event, life on Earth seems to have recovered quite nicely, albeit with much different floral and faunal components. However, for many millions of years right after the P/T catastrophe, complex life itself was in jeopardy. Some plant and animal species survived through the sheer luck of being in the right place at the right time and with the right abilities. Unfortunately, most species weren't so lucky. Their sole remnants are those preserved in the fossil record. As time passed, the species that did survive adapted to their new environments, and even the survivors were eventually lost.

At the time of the P/T catastrophe, the oxygen levels dropped to a maximum of 15% (Berner 2006) and likely less than 13% (Uhl, 2008) in Earth's atmosphere compared to our present 21%. Humans would find it very difficult to survive under those conditions and would undoubtedly find breathing at any elevation much higher than sea level virtually impossible were similar oxygen levels present today. An interesting piece of information is that so much oxygen is used up near large cities via the burning of fossil fuels that local atmospheric oxygen levels often fall significantly. It is only because of oxygen being replenished by photosynthesis occurring primarily in Earth's

Figure 23. Permian chert is made from the remains of sponges, brachiopods, and other marine organisms.

oceans that this effect hasn't become a major health problem. Should conditions in the oceans jeopardize the existence of such photosynthesis by phytoplankton, the lowering atmospheric oxygen concentrations will become an immediate and serious problem. This situation is double-edged: the same phytoplankton are responsible for breaking down the bulk of the Earth's atmospheric carbon dioxide.

Some of the effects of the P/T catastrophe on Earth's oceans are still readily observable today. At the end of the Permian Period, carbonate "coral" reefs were being built worldwide. These reefs were being built not only by many groups of corals, but also by sponges, brachiopods, and other marine organisms. An interesting group of animals called radiolarians produce silicate body structures rather than carbonate. Sponges also contain silicate structures. These remains sink to the ocean floor and form chert. Thick layers of cherty, silica-rich sediments covered the Permian ocean bottoms. However, for over 10 million years after the P/T event, the fossil record contains no evidence of either chert or carbonate reef structures. One could argue that the animals survived the catastrophe, but that the ocean chemistry at that time prevented their preservation in the fossil record. However, since cherts are largely unaffected by most conditions including increased acidity, the most logical interpretation is that the Permian reef-building animals died out.

Luckily, some reef-building organisms must have survived and, after millions of years, were again able to repopulate the oceans. The record of reef-building does show virtually an entirely different group of organisms involved between the Permian and Triassic reef structures (Pruss and Bottjer 2005). The Triassic reef builders were not survivors from the previous lineages – they were distantly related organisms that modified and adapted to fit a new life mode for their lineages. The Earth has still not recovered its reef-making ability to the levels seen in the Permian.

Of all the time represented in the rock and fossil records, no other single point records such a massive change in climate and biosphere interactions. Unfortunately, the most catastrophic event is often ignored due to the longer-term events leading up to the catastrophe. Indeed, many scientists still refer to the P/T as a long-term extinction event, where terrestrial volcanics gradually poisoned the atmosphere and oceans (e. g. Beerling et al 2007). While this is true for the period of time leading up to the cataclysm, it definitely does not explain the cataclysm itself. Only one scenario has been proposed that fits the current data and explains the catastrophic rise in temperature and geologically instantaneous changes in ocean and atmospheric chemistry – the massive and catastrophic release of methane from the ocean floor.

Methane Clathrate

As stated earlier in this volume, methane clathrate, methane hydrate, gas hydrate, or methane ice, as it is variously called, is a substance that forms on continental shelves beneath the oceans and under tundra landscapes (Buffett and Archer 2004). Currently, and likely at the end of the Permian Period as well, over half the Earth's carbon resources were/are tied up in this form. Unfortunately, this form is only metastable. As long as its environment is kept cool, the methane clathrate stays frozen and keeps the methane from releasing into the biosphere. However, should temperatures rise, the ice breaks

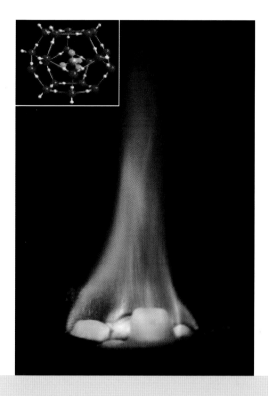

Figure 24. Burning methane as it is being released from a piece of methane clathrate. (photo courtesy Wikipedia)

down and huge volumes of methane gas are released (Hill et al. 2006).

Methane gas is poisonous to most marine organisms. Its diffusion into ocean waters both removes the oxygen most marine organisms need to survive and reacts adversely with organisms' metabolisms. In the atmosphere, it is capable of capturing 20 times the amount of solar energy as a similar amount of CO_2 and also of causing regional, or possibly even global, flash fires. Perhaps the worst aspect is that, as it breaks down, for every molecule of methane removed, a molecule of CO_2 is produced. It has, therefore, both short term and long term effects.

Fortunately for marine life at the moment, most of the methane currently being released from clathrates forms bubbles that grow progressively larger as they rise to the ocean surface. These bubbles lose very little of their content to diffusion into the ocean – over 90% of the released methane rises to the surface and escapes into the atmosphere (UC Santa Barbara press release, 2006). This process is currently happening near Bermuda, and these bubbles are likely the cause of some of the mysterious disappearances that have happened in that region. A bubble released from a thousand feet under the ocean would expand over 160 times its original size by the time it reached the surface. Thus, a bubble 3 feet in diameter at depth would be over 450 feet in diameter when it reached the surface! Even a large ship would likely not survive falling to the bottom of such a bubble if it rose underneath the ship. Likewise, such a bubble released into the atmosphere would stall aircraft engines and possibly render flight crews unconscious (unless it was already partially mixed with atmospheric oxygen, where it would cause an explosion!).

At the time of the P/T catastrophe spike, the temperature of the ocean floor apparently reached the point where significant portions of the clathrate began to melt. Chemical traces in rocks of that age indicate that this trigger point occurred somewhere between when the Earth's atmosphere had two and four times the current atmosphere's CO_2 content (Retallack 2005). It appears that the circulation of cool/cold water at depth in the oceans significantly diminished, likely due to the loss of cold water flowing off melting polar ice caps. This was, in turn, likely due to the ice caps becoming mostly or completely melted as increased CO_2 levels gradually increased global average temperatures. Once this circulation vanished, warmer conditions prevailed in the ocean depths. The ocean floor water temperature rose, melting the methane ice. Once significant amounts of methane were added to Earth's atmosphere, much more solar energy was trapped, further increasing the ocean temperatures and speeding the clathrate melting process. At least one researcher has suggested that this process went from start to finish in as little as 15 years (Dorritie, 2007).

Regardless of the actual time it took for the methane clathrate collapse to occur, the effects are well-documented. Globally the temperatures spiked at least 5 degrees Celsius (9 degrees Fahrenheit) over the additional 5 degrees Celsius that temperatures had gradually raised earlier. A portion of the methane released earlier into the atmosphere was returned to the oceans via rain entrapment, and this methane became a significant factor in the sterilization of life from Earth's oceans at that time. As the methane reacted with oxygen (read, "burned"), the oxygen level in the atmosphere was reduced by half (Retallack 2005). Over half of all land-based genera and over 90% of all marine genera were lost. Since a genus can survive from a few mating pairs, this means that it is likely that well over 99% of all life on Earth died at that time. To lose large populations such as happened during that catastrophic event, the planet had to have been globally inhospitable to life. This was not a case of accelerated extinction as we are seeing at present; this was a case of the entire planet being virtually uninhabitable.

Today, we are seeing conditions on Earth almost identical to those present at the end of the Permian Period. We currently have receding glaciers, melting polar ice caps, methane release, and CO_2 buildup. And, most importantly, we have the mechanism in place to continue the CO_2 buildup past the "twice current levels" situation that may have triggered the P/T event. If we witness a major volcanic eruption such as is recorded in the rock record but larger than those seen in human history, the added CO_2 levels in the atmosphere could trigger a methane clathrate release. However, even without such a trigger, unless human practices change, the requisite amount of CO_2 will still be added to our atmosphere by the end of this century through the burning of fossil fuels, and primarily through coal-fired electrical generation.

This was a case of the entire planet being virtually uninhabitable.

Chapter Six

Earth's Carbon Balance

Of all the topics covered in this volume, this is undoubtedly the most controversial. The reason is that
this concept is largely based on my own derivation and
synthesis. However, as has been presented in previous
chapters, global temperatures throughout Earth's history show a strong direct correlation with the amount
of carbon circulating through the biosphere, and it is
well-documented that the majority of Earth's energy-
trapping mechanisms involve carbon. This concept can
be summarized as, "with factors such as solar energy
output and Earth's albedo remaining constant, Earth's
global temperature balance is regulated by the amount
of carbon in the biosphere balanced against what is un-
available/sequestered." The balance of this chapter is

dedicated to an explanation of why this is the case and its ramifications and practical applications to modern society. Some of the more important points of this discussion are highlighted.

Natural or Human-Caused?

One of the greatest controversies concerning climate change is the question, "Is the current climate change a natural or human-caused event?" The answer is yes.

The Earth, at some point, would follow the same pattern as in the past. With or without human activity, the Earth will cool, build up clathrate, slowly warm for some reason until the clathrate begins to melt, then catastrophically warm due to acidification/sterilization of the oceans allowing buildup of carbon-based gases in the atmosphere. This cycle has repeated in the past, and the cycle will continue to repeat in the future.

The timing of these events is, without doubt, being affected by human activity. What would normally have happened thousands, if not millions, of years from now will likely happen dozens to hundreds of years from now instead. The trigger, as per all the information in the rock and fossil records, is a buildup of CO_2 in Earth's atmosphere allowing a warming that, in turn, melts the polar ice or causes a die-off of the ocean phytoplankton. Once these temperature-limiting factors are gone, the methane clathrate is released, and the Earth warms rapidly and catastrophically until a new equilibrium is reached. Humans, through their practice of increasing the amount of CO_2 in Earth's atmosphere, are hastening this process.

The catastrophic event is part of an ongoing natural process. However, that process routinely wipes out most life on Earth. While it is, indeed, a natural process, humans have the right, and perhaps the means, to prevent their demise through this natural process. In reality, modifying this event is no different than building dams and levees along a major river to prevent floods from killing people downstream. The only difference is the scale. This is a global event, and modifying it will take a concerted global effort.

The Human Component

Throughout this book, I have talked about various interactions between human activity and global climate. During the past few years, the IPCC and others have documented these interactions in detail. No objective analysis of the data would allow any other conclusion but that human activities are having an effect on global climate. However, world leaders, scientists, and the public are seemingly bogged down in wrangling over the questions of whether the effects are positive or negative, how significant the effects are, and what, if any, are solutions. Because of these wranglings and conflicting claims, it is easy for humans in general to ignore the whole issue and hope it goes away. However, should paleontologic events repeat, our grandchildren would hate us to their dying days for taking that path when we had the chance to prevent the catastrophe. If we truly wish to "save the planet," we first have to overcome basic human nature.

What, exactly, is the problem?

Ordinarily, it would not be a problem for the Earth to warm. In general, species survive and thrive well in warmer climates, and a warmer Earth would open new temperate and polar areas to habitation. What would be lost at lower latitudes and elevations would be gained at higher latitudes and elevations.

However, catastrophic warming caused by massive release of methane would be a possibly insurmountable problem for the human species. Based on the problems previously described concerning the volume and lack of long-term stability of Earth's methane clathrate deposits, the option of "allowing nature to take its course" would likely mean the extinction of the human species. I personally do not accept this as a viable option.

The Earth must be kept cool enough for long enough that we can convert the existing methane clathrate into more stable forms.

Carbon, nitrogen, and oxygen are elements with 6, 7, and 8 protons, respectively. As such, they cannot be broken down into simpler forms. Except in nuclear reactions, they can neither be created nor destroyed. They are (except in rare instances) too heavy to escape Earth's gravity. Of these three,

Figure 25. While many sources of methane have been blamed for global temperature increases, methane clathrate is by far the most significant.

only carbon exists in the solid state at the temperatures found on (and near) Earth's surface. It is interesting that the most common nitrogen compound is formed with its oxygen neighbor to create a compound that is very good at trapping solar energy in Earth's atmosphere. However, it is carbon in its various compounds that is most likely to be involved with energy capture and transport.

Since the carbon atom is responsible for trapping solar energy in so many different ways, it is imperative to establish some method for differentiating between carbon that is not active in the energy capture processes and carbon that is. Obviously, carbon that is buried deep in the ground is unlikely to be involved with the retention of solar radiation. Diamond and graphite are also unlikely to become involved even if they are once again brought to the Earth's surface. On the other hand, the role of methane and carbon dioxide in trapping solar energy in Earth's atmosphere is well documented. But what about the carbon in plants and animals? Or the carbon in our soils? And what are the main processes whereby carbon is released or sequestered? Is there an easy way to separate carbon that isn't involved with global warming from that which is? Luckily for us, there is.

Carbon exists in several different allotropic forms (same element composition but molecules arranged in different ways) and also in a myriad of compounds. Amorphous carbon, graphite, and diamond are the most common of the carbon allotropes. Of these carbon forms, diamond has the most tightly packed atoms. It is generally formed under tremendous heat and

The only process that breaks down CO_2 and releases O_2 to any significant degree is Photosynthesis.

pressure, and it has only been in the last 50 years that manmade diamonds have been produced in any quantity.

Graphite is unusual. It is a very dense form of carbon that is usually formed through metamorphic processes, but can also be formed in hydro-thermal situations (Coverdell, pers. comm., 2010). It requires a great deal of heat and pressure in its creation, and natural deposits are rare except in areas where relatively pure carbon deposits were subjected to such.

Amorphous carbon is a black substance found in lamp black, charcoal, coal itself, and in microscopic particles throughout the biosphere. Pure amorphous carbon is primarily formed when long-chain hydrocarbon compounds are subjected to heat and pressure without sufficient oxygen present.

Coal is not an allotrope of carbon. However, much of Earth's carbon is tied up in the form of coal. Coal is actually a mixture of various hydrocarbon compounds and chemical arrangements and mixed with various impurities present when the deposit was created. A purification of these deposits occurs when they are subjected to heat and pressure, and elements other than carbon are preferentially driven off during this process. The lowest grade of carbon deposition in this process is peat and similar organic deposition. As those deposits are subjected to more heat and pressure and for longer periods, the process produces lignite, bituminous, and anthracite coals, respectively.

These forms of coal range in carbon content from 60% or so for lignite to over 90% for anthracite. Most coal deposits are 80% to 90% carbon, and the coal molecules are very similar to carbon itself in chemical properties and stability. For these reasons we will consider coal and amorphous carbon in the same manner for the remainder of this work.

Carbon also exists in many other forms as components of organic compounds. Several of these have been the subject of discussions earlier in this work, and they are integral to Earth's energy transport system. Indeed, organic compounds are usually found as part of the biosphere. Only methane clathrate, petroleum, natural gas, coals, tars, and asphalt are commonly sequestered carbon compounds.

Carbon in its pure form and carbon compounds that are deeply buried or transformed into inert allotropes or stable compounds do not interact in the global warming process. All other carbon is transient in the system and thus may contribute, at least intermittently, to energy capture and global warming.

For instance, grass is a green plant that builds its structure by taking carbon dioxide from the atmosphere, releasing the oxygen, and using the

Figure 26. However beautiful and diverse they may be, rainforests have a negligible effect on global climate.

carbon to form complex hydrocarbon molecules. After only a few days of growth, a cow can eat that grass and, in the process of digestion, release significant amounts of the carbon trapped in the plant structure as methane gas (Thorpe 2009). The rest of the "grass carbon" goes either to the building of the cow's body or is passed through as exhaled carbon dioxide or as feces. The feces eventually rot and continue the cycle, as does, eventually, the cow's body. Grass that is not eaten is eventually either rotted, burned, or otherwise recycled. Thus, even though the plant removed the carbon from the atmosphere, the end result is that the atmosphere receives back as much or more energy trapping capability. This particular cycle does nothing to change the energy entrapment pattern even though the plant initially removed a significant amount carbon from the atmosphere.

Similarly, trees do remove atmospheric carbon to build their structures, but no tree lasts forever. Most trees end their lives in forest fires only a few decades after their creation, and the stored carbon is released back into the atmosphere. While trees remove atmospheric carbon and other pollutants, they also give off methane when they rot or are eaten by termites. While they provide shade, they also absorb significant amounts of solar energy. In all, trees do little, if anything, over the long term to regulate global warming or lessen the atmospheric greenhouse effect unless they form coal.

Although plants are key to the carbon sequestration process, in most cases the carbon escapes and continues to recycle through the system. Merely growing additional plants will not solve the problem of too many carbon atoms in the biosphere.

Figure 27. Photosynthesizing organisms in the sea, such as algae, are a significant part of the carbon recycling process.

So, how does one remove atmospheric/biospheric carbon in a semi-permanent manner? The only process that breaks down carbon dioxide and releases oxygen to any significant degree in Earth's system is photosynthesis. Other processes that break down carbon dioxide and sequester carbon, such as the formation of carbonate rocks, require the removal of oxygen as well as carbon from the atmosphere. Through photosynthesis, the Sun's energy is used to strip carbon atoms from carbon dioxide, and then more energy is used to create the complex hydrocarbons and organic molecules in the bodies of photosynthesizing organisms. In turn, these organisms are eaten by other organisms that cannot photosynthesize. Once there, they, in turn, are eaten by other organisms.

A small percentage of terrestrial organisms in this cycle are buried so deeply that the carbon becomes trapped underground when the bodies decompose. This burial occurs commonly in peat bogs. However, most of the photosynthesizing organisms on this planet do not live on land, but rather, in the sea. They are the tiny phytoplankton -seaweed, cyanobacteria, algae, and other photosynthesizing organisms that inhabit Earth's oceans. These organisms inhabit the upper 600+ feet (200 meters) of the ocean where sunlight can penetrate. Most of the oil and gas reserves on this planet can be accounted for by the burial of these marine or lacustrine organisms and their feces at the bottoms of bodies of water.

It was long thought that these organisms almost always sank to the bottom of the ocean and once there, produced the carbon ooze that later would become oil and natural gas. Recent studies have shown, however, that much of the carbon remains in the upper ocean layers. Non-photosynthesizing organisms eat the photosynthetic ones, and are, in turn, eaten by other organisms. According to these studies, only the largest of carcasses and fecal pellets actually sink to the bottom. And, once there, they provide food for a host of other organisms living in the realm of the deep.

Decomposition of dead organisms and their waste produces methane. This methane is then bubbled back to the ocean surface, trapped by the formation of clathrates, or absorbed into the ocean water itself. It is only in the formation of methane clathrate (see chapter 4, methane section) or by the burial of organics before their conversion to methane that this carbon is trapped for indefinite amounts of time. In order for carbon ooze to form, burial must occur rapidly enough to prevent much of the decomposition, and this seldom happens. Only when underwater landslides or heavy dust falls provide enough sediment to bury the remains deeply does decomposition cease prematurely. It is primarily the lack of free oxygen within those deep sediments that allows long chain hydrocarbons to accumulate.

Thus, it is an ongoing mystery how, exactly, crude oil is formed. The above-described recent data would make the generally accepted scenario unlikely. There just would not be enough of a steady accumulation of organisms sinking to the bottom and creating a carbon ooze. Another peculiar event is that some oil wells are continuing to produce high volumes of oil long after their calculated subterranean reservoirs should have been depleted. There is no way to explain these anomalies using the current model of crude oil creation.

Figure 28. It is an ongoing mystery how, exactly, crude oil is formed.

The point of this discussion is that the semi-permanent sequestering of carbon is a rare event. Once carbon is in place in the biosphere, it is recycled again and again. It is the rare event that carries a carbon atom into a situation where it no longer interacts. Coal beds, oil and gas fields, and even tar sands are some of these rare events that took millions of years to create. Unfortunately, it has only taken us a few decades to seriously diminish these deposits.

Another misunderstood model concerns the Earth's rainforests and their effects on atmospheric carbon content. The rainforest has long been touted by ecologists and environmentalists as one of the most important regions on Earth for cleansing the atmosphere of CO_2 and enriching its oxygen content. The rainforest has also been long recognized as providing cooling and shade and thus, by inference, preventing the desert heat found elsewhere at similar latitudes.

Unfortunately, these suppositions have not held up well in light of current research. While primary rainforest does convert carbon dioxide to cellulose and other hydrocarbons plus free oxygen, the CO_2 is returned to the environment when the trees burn or rot. Unlike peat bogs, there is no evidence of fossilization or burial of carbon in current rainforests. In addition, satellite imaging has recently revealed that tropical rainforests are the world's largest single source of atmospheric methane (Miller et al 2007).

Rainforests have also been touted as great structures for raising albedo and thus reducing the greenhouse effect. Rainforests do reflect significantly more light than oceans - approximately 25% of the solar energy falling on them, but deserts reflect 40%, and concrete reflects 55%. So, although the rainforests reflect more energy than the oceans, they reflect significantly less than the deserts that could replace them. Rainforests also increase the humidity in their regions. As mentioned earlier, atmospheric water vapor (humidity) traps more of the Sun's energy than all other greenhouse gases combined.

Less than 10% of Earth's terrestrial surface is covered by rainforest, 10% is covered by glaciers, and only 1/5 of the Earth's surface is terrestrial to begin with. Thus, rainforests occupy only about 1% of the surface of the Earth. While rainforests have a number of beneficial aspects and are well worth protecting, their influence on global climate is minimal.

In the grand scheme of things, rainforests have a negligible effect on global climate.

The simplest route to permanent carbon removal from the biosphere is the production of carbonate by marine organisms. In both numbers and tonnage, vast quantities of life forms that produce carbonate body parts exist in Earth's oceans. These include individual shells such as those of clams, snails, and many types of microscopic animals and also colonial structures such as coral reefs. This carbonate does remove large volumes of carbon from the biosphere. Unfortunately, this process also involves removing double or triple the number of oxygen atoms for each carbon atom removed. The most common permanent sequester of carbon is through the manufacture of calcium or potassium carbonate as shells for various marine micro- and macro- organisms. However, these structures sequester carbon as carbonate (CO_3). Unfortunately, this means that sequestering carbon in this manner uses up more atmospheric oxygen than CO_2.

Although coral reefs and other limestone and dolomite features are produced by aquatic organisms and are composed partially of carbon, three or more oxygen atoms are taken out of our atmosphere for each carbon atom sequestered in this manner.

In light of the information presented thus far, it is obvious that carbon atoms are commonly found in organic compounds and only more rarely in a pure form. They react often, are very important in the solar energy capture process, and are only easily sequestered/buried in the form of hydrocarbons. Once carbon atoms are buried deeply enough that oxygen cannot be supplied in quantities to support rapid oxidation (burning), they generally become inert. Likewise, once carbon atoms are forced into their graphite or diamond forms, they also tend to remain stable over the long term. Certain other compounds, notably silicon carbide (one of the most common abrasives used for sandpapers and grinders) and carbon dioxide, also remain stable. However, inert carbon compounds are relatively few. Volatile, reactive carbon compounds are much more common, and even the relatively non-reactive carbon dioxide is still a major energy entrapment molecule in our atmosphere.

As mentioned earlier, only three factors have been shown to be consistently correlated with average global temperatures: solar output, the Earth's albedo, and the Earth's atmospheric composition. Obviously regulating solar output is not possible. Should we desire to stop the increase in global average temperatures, we need to find ways to further limit the amount of energy trapped from the sun. We, theoretically, could change Earth's albedo - turn vast areas of our world into lighter-colored surfaces (converting the remaining rain-forests into desert comes to mind) and thus increase reflectivity. However, changing Earth's albedo is a complex process, and we would have no way to estimate the effect until after changes were made. Such experiments could inadvertently turn catastrophic. We could easily trigger the next ice age if we raised the albedo too high, trigger methane release if we left it too low. And, within our present level of understanding, changing the Earth's albedo could not be accomplished precisely enough or changed rapidly enough to maintain the delicate balance required.

Far less risky and far more predictable is the process of changing Earth's atmospheric composition. We already have decades of cause-and-effect data and we know in sufficient detail which atmospheric compounds are desirable and which are to be avoided. Many energy-trapping carbon compounds are present in the atmosphere, and we have the technology to understand and to, at least partially, regulate their concentrations.

Our best chance to regulate atmospheric composition is to regulate energy-trapping atmospheric carbon compounds. However, it does no good to suck existing carbon dioxide out of the atmosphere if it will just be replaced by other carbon dioxide. In order for real change to take place, the total amount of available carbon in the biosphere must be limited. Were we, through several activities currently possible, to sequester as much carbon in the form of charcoal as we are currently removing in the form of coal and were to eliminate the removal of sequestered coal, we could lower available atmospheric carbon compound concentrations considerably. We could then make room in Earth's global energy budget to convert existing clathrate deposits into other compounds, possibly even using the energy from those processes for our energy needs.

Only by removing the methane threat while simultaneously preventing additional warming are we going to defuse the clathrate timebomb. This requires both prongs of a 2-prong approach. We must "burn off" the existing clathrate, but we must sequester as much carbon as we release. Otherwise our actions will cause the Earth to warm and the methane to release catastrophically.

Luckily, we are warned. We have time, and we have technology.

This problem is not insurmountable.

Luckily, we are warned. We have time, and we have technology.

Chapter Seven

**Current Climate
Change Issues**

Can This Continue?

Current climate regulation initiatives are pretty straight-forward. The following description from climate-leaders.org (a climate initiative in the country of India and part-nered with the U.K.) provides the most concise summary of current thought on the topic:

The challenge of preventing climate change is formi-dable but not insurmountable. To avoid a dangerous lev-el of climate change, defined by the European Union (EU) as an increase in the global mean surface temperature of 2°C or more above pre-industrial levels, developed coun-tries will be required to stabilize and then cut their cur-rent greenhouse gas (GHG) emissions between 60 and 80 percent by 2050. Developing countries will have to stabilize their emissions while finding new, low carbon pathways to development. Timing is also critical: the longer action is delayed the steeper the reduction track will become and the harder climate change will be to manage.

Put simply, there are three ways to prevent climate change:

> *1. Reduce emissions through efficiency – using less*
>
> *2. Reduce emissions through substitution – using something else which is less harmful*
>
> *3. Reduce emissions through sequestration – using something to capture and store emissions*

Let us examine the issues raised by each of these points:

The problems with using increases in efficiency to prevent climate change

While it is very important to conserve, the simple fact that our population is doubling every 30 years or so argues that, in the long term, we won't be able to conserve as fast as we populate. Right now, most of us use fluorescent light bulbs in our houses that use ¼ of the energy when compared to the incandescent bulbs primarily used in the past. However, we are also lighting twice as many houses and using more bulbs to do so. In all, efficiency in this area has kept electrical usage levels for lighting nearly the same for the last few years; it has not reduced electrical demand. The only way we would show a decrease would be if we actually lowered our standard of living – did without many of the electricity-consuming amenities we now consider necessities.

Even if we don't take into consideration the projected population growth, other factors also limit the effectiveness of increasing efficiency as an answer to climate regulation. First of all, since climate regulation is a vast, complex, largely natural process, the interaction of humans is only minimally responsible for the process in any case. One volcanic event can still cause more disruptions to Earth's climate than an entire year's cumulative human activities. Conservation could account for, at most, a fifth of our climate-affecting activities which, in turn, accounts for only a small portion of the largely natural process. Twenty percent of twenty percent is only four percent. The Earth's processes will still be driven by 96% of the force originally available. While this may slow the process slightly, it could never halt it.

Perhaps the most common misperception is that the Earth is static. If we are only talking about a system where human activity is directly attributable to observed climate change, then much of what has been said about conservation makes sense. As long as we have this picture of the Earth's living populations as being healthy and self-sustaining except for when humans interfere, human interference obviously becomes the visible cause when populations decline and groups go extinct. These viewpoints, while prevalent, have no basis in fact. All the data from a paleontological perspective shows a highly variable planet on which life has nearly been wiped from its surface several times. We observe our pollution, we observe extinction, and we immediately infer cause and effect. While in specific instances this can be shown to be the case, in the vast majority of instances it is not. The Earth will not continue as we know it regardless of how much we conserve. Ultimately,

All the data shows
a highly variable
planet on which
life has nearly been
wiped from its surface
several times.

the predicted catastrophic event would have happened even if humans had never set foot on the planet. We have merely speeded up the timetable.

The problems with using substitution to prevent climate change

Primarily, when one looks at this issue, one thinks of things like paper versus plastic. We have all heard how plastics don't break down and are difficult to recycle, so we should use paper, then recycle it (or at least recycle the plastic). However, what effect do these activities have? If you look at the costs in terms of fuel burned and electricity consumed, it may be more costly to the environment for an individual or household to recycle than not. On the other hand, recycling at the landfill site can be very efficient, but it is seldom implemented at that point. In my opinion, more effort should be expended toward streamlining this process with the goal of actual environmental benefit rather than a "feel-good" environmentalism of blue plastic buckets and bins (plastic buckets for recycling? Is there something ironic about that?).

Substitution is actually the heart of what can help with the current global crisis. Yet it is often implemented poorly or not at all. Substitution at the energy producing level is currently not happening. As stated throughout this work, coal-fired electrical generation is the largest single climate-affecting human activity, yet the number of coal-fired plants is increasing rather than decreasing. The fact that coal provides most of the backup for solar and wind generation, coupled with its abundance and its lack of perceived environmental drawbacks make it the mainstay for electrical generation in many countries. Even in the United States, current policies often prevent available hydroelectricity from being added to the grid if commercially produced coal- or gas-generated electricity is available. By statute, in this instance we must pollute to the max before we can add from non-polluting sources.

Substitution of electricity for gasoline in automobiles is gaining wide support. New technologies, coupled with perceptions of "pollution free" driving are fueling this movement. However, in the U.S., cars plugged into the grid are getting nearly 2/3 of their energy from burning fossil fuels. Additionally, energy and other costs and availabilities associated with production of the rare raw materials needed to create efficient electrical transportation currently does not allow for a major conversion. Although we have enough re-

Figure 29. Coal-fired electrical generation is the largest single climate-affecting human activity.

sources for some electric cars, we don't have enough to supply even a majority of our transportation needs. Charging vehicles from the electric grid itself may also strain the current system past capacity. No hard numbers exist, but most indicators point to an end result where we would be merely trading using gasoline for using coal in roughly the same amounts to power our future transportation needs.

Substituting hydrogen for gasoline to power some, or all, of the various aspects of transportation has been cited often as a very efficient means of reducing emissions. Other such substitutions involving non-greenhouse gas byproducts have been proposed as well. Unfortunately, virtually all these products are years away from effective implementation, and most require significantly more energy and higher costs of operations than those currently employed. By the time any of these options are commercially available, it will be too late to have much beneficial effect on the climate problem.

Transportation is a huge concern. As Earth's population both increases in numbers and becomes more mobile, passenger transportation increases exponentially. As population increases, the amount of transportation of food and merchandise also increases dramatically. In all, we are facing a marked increase in transportation demands in the next generation. To meet these demands, we have only made the current system more efficient; we have not created any viable alternatives. Without a viable, non–fossil-fuel based alternative, transportation will overtake electrical generation as the most critical climate problem in the near future.

Figure 30. Transportation will overtake electrical generation as the most critical climate problem in the near future.

Problems with using emission sequestration to prevent climate change

Sequestration – the permanent entrapment and containment of harmful emissions – has been touted as the ultimate means for reducing the effects of human activities on Earth's climate. In principle, sequestration of greenhouse gases is the ideal solution. The reality of the situation has little to do with this ideal.

First of all, we have no technology available to sequester large quantities of greenhouse gases. Most proposals call for the pumping of these gases deep underground. However, there are no guarantees that a leak won't develop from such an area, and there is no known method for sealing such a leak. Were such a leak to occur, the release, like that of a volcanic eruption, could actually trigger the event we are trying to prevent.

The majority of strong advocates for such gas sequestration are oil producers who could use such a system to pressurize low production oil fields. By such pressurization, they could then return marginal fields to full production. One cannot help but see the irony in sequestering harmful fossil fuel byproducts in such a manner that more fossil fuels are able to be produced.

Carbon is a very reactive molecule, and there exist few compounds containing carbon that will not react over time. The most stable carbon forms are those in coal, graphite, and diamond. These forms can exist in a wide variety of temperature, humidity, and pH conditions, making them ideal for carbon sequestration. Unfortunately we are using the bulk of Earth's known outcrops of these – especially coal - for the opposite purpose.

Other carbon compounds, including a variety of plastics, silicon carbide, carbonates in rocks, and carbon dioxide, are also quite stable. In fact, the stability of carbon dioxide in Earth's atmosphere is the single most important temperature regulating factor in Earth's climate. As discussed earlier, sequestering any gas is problematic, and yet we do need to lessen the amount of CO_2 in the biosphere. We need to avoid the problem of gas sequestration by converting carbon dioxide into other stable forms that are easier to sequester.

Plastics vary widely, and efficiently sorting environmentally stable plastics from others would be difficult. Producing such stable plastics for the simple purpose of removing carbon molecules from the biosphere would be cost-prohibitive. Likewise, silicon carbide is a wonderful carbon-sequestering compound, since it bonds carbon with silica and releases oxygen from both silica sand and carbon dioxide in the process. However, it is most commonly produced at extremely high temperatures (greater than 1000 degrees Celsius/1800 degrees Fahrenheit) and the temperature maintained for many hours (Sudarshan and Maximenko 2006). Such input of energy makes using these materials impractical for large-scale carbon sequestration.

Of all potential carbon sequestration options, only carbon in the form of coal or charcoal is cheap, easy to produce, and easy to store indefinitely.

From the above descriptions, it is obvious that the current efforts proposed through the Kyoto Protocol and other climate change prevention efforts fall short of providing working solutions to the climate change problem. Compounding this problem is the fact that certain major governments, including that of the United States, have refused to either agree to the Kyoto Protocol or set up mandatory greenhouse emission restrictions of their own. In the following sections we will examine some of the problems with U.S. efforts to curb climate change and some lack of understanding of the issues themselves.

Greenhouse Gas Reduction

Currently, the United States has no mandatory policy in place for lowering greenhouse gas emissions. It does, through the EPA, provide voluntary guidelines, tax credits for "green" practices, fines for polluting past certain levels, and the options of "cap and trade" for balancing polluting with greening practices. However, these policies are currently based on arbitrarily set levels of pollution, and there is little data to support these levels as being below the future catastrophe trigger point. In fact, the United States is criticized worldwide for its allowance of more environmental pollutants than the Kyoto Protocol allows.

Unfortunately, at the heart of the worldwide climate issue is the fact that no regulatory body anywhere makes a distinction between emissions from currently sequestered carbon and emissions from carbon already present and active in the biosphere. The idea that producing CO_2 is a bad thing has permeated human understanding worldwide, and this understanding has led other countries including Russia to adopt strict standards on the AMOUNT of CO_2 produced (Kyoto Protocol). While laudable, these efforts are based on misunderstanding the problem itself.

If one looks at the greenhouse gas issue as one of carbon balance – the amount of carbon interacting in our biosphere versus that sequestered--one can immediately see the problem with current government policies. Even if we reduce CO_2 emissions by half, if that half comes from carbon not previously in the system, we have actually INCREASED the greenhouse gas levels in our biosphere.

Conversely, if we double our CO_2 output, but all the CO_2 comes from living plants and other circulating carbon sources, we have NO NET GAIN in total greenhouse potential of the planet over the long term (please note the words "Long Term". At present, we are in a situation where even short-term increases in CO_2 are dangerous and should not be allowed.). The current policies and procedures being advocated to reduce greenhouse gas emissions miss this point entirely. Because of this, billions of dollars will be spent fighting climate change, and the CO_2 levels will continue to increase.

Currently the U.S. has no mandatory policy in place for lowering greenhouse gas emissions.

Alternative Energy Sources and Energy Credits

If you live in the U. S. Midwest, and increasingly elsewhere in the nation, you are likely to see windmills if you drive very far. Because of current energy policies, it is profitable for the larger greenhouse gas producers to invest heavily in technologies that provide electricity without pollution. Windmills have become the product of choice because they do not pollute, they don't interfere with waterways, and they don't involve nuclear reactions. Even though windmill-produced electricity costs 10 times more than hydroelectricity, paying for the construction of a windmill can oftentimes be profitable due to credit against serious taxes and fines levied against an air polluter. The downside to this practice is that the windmills do require maintenance, and when there is no longer government incentives to maintain these structures, they tend to fall into disrepair.

No "green" technology currently used on a significant scale, other than hydroelectricity, is capable of producing dependable energy. This includes both wind and solar electrical sources. If the wind doesn't blow, or if the day is cloudy, electricity isn't produced as necessary. In order to make these technologies functional, there has to be an alternate source that can immediately supply power when wind or solar sources can't. While hydroelectricity, geothermal electricity, and fossil-fuel electricity could all theoretically fill this need, the alternative source of choice is almost invariably coal-fired electrical generation.

Hydroelectricity does have its own unique problems. Because hydroelectricity is generated using dams, fish migrations are often affected, scenic areas and productive croplands are flooded, and natural flooding cycles are disrupted. Also, deeper water provides different ecologic and environmental effects than shallow, so the pre-dam environment is changed. For these reasons, permits for new structures are difficult, if not impossible, to obtain, and the amount of hydroelectricity being produced in the U.S. has actually declined in recent years.

Perhaps the most significant problem with hydroelectric generation is that the U. S. Government claims all waterways and owns most of the dams. Since the government is not allowed to compete with the private sector, government-produced electricity cannot be introduced into a grid where its addition is at the cost of lost profit to a private entity. Thus, most dams nowadays provide only backup power to the grid. Also, most dams were installed with flood control or shipping in mind, and electrical generation was added as an afterthought. For instance, Fort Peck Dam in Montana sees peak flows of over 50,000 cubic feet per second, but the electrical generation plant is capable of accepting less than 10% of the dam's peak flow (pers. comm., McMurry, 2010).

What all the above factors add up to is that, in order to make wind and solar generation work, we are actually burning fossil fuels. We have to have an energy source that is available at a moment's notice; one that is owned by the private sector, and one that the general public does not object to. While it has been shown that coal-fired electrical generation puts more radioactivity back into the biosphere than a nuclear power plant (Gabbard, 1993), the public perception is still, "coal=good, nuclear=bad." And, coal has one char-acteristic that often outweighs other considerations: We still have abundant coal reserves within the United States.

So long as the U. S. and China have abundant coal reserves, there will likely be no effort made to replace coal-fired electrical generation in those countries. To do so would be both economically and politically unproduc-tive in the short term. It is a sad reality, but no leader is going to sacrifice current political security and profitability for prevention of an event that likely won't happen in his or her lifetime. Only through united demands by the majority of the world population will that attitude change.

In order to make wind and solar generation work, we are actually burning fossil fuels.

Chapter Eight

What Can We Do?

Our Impact

You have had a brief glimpse into the complex interactions that we collectively call climate, and I hope you can see that the Earth is a very large, complex system. It is also a mostly closed system, as no significant amounts of matter are either being removed or brought in. Hydrogen, the very lightest of elements, is the only one that escapes Earth's gravity in any appreciable amounts (an estimated 3 kilograms of hydrogen are lost per second vs. 50 grams of Helium, the next lightest gas). If we model the Earth's biosphere as a closed system when it comes to carbon compounds, then it is obvious that, since carbon atoms form part of most greenhouse gases, we should be looking for ways to prevent carbon from forming these compounds.

The reality is that the Earth is so large and complex, it is doubtful that we will ever be able to control these reactions and atmospheric components. Reports from the Mauna Loa reporting site have shown a rapid, steady, and ultimately predictable rise in atmospheric CO_2 over the past 50 years (Tans, 2010). This approximately 2 parts per million rise in atmospheric CO_2 per year has already caused observable global increases in temperature.

However, life on Earth did very well when the Earth was warm, and should the Earth transform into a warmer place once again, life would continue to do well. The only problem is that, in getting from point A to point B, the Earth tends to become very inhospitable. 10 million years after the Permian/Triassic event, the Earth was stable, temperatures were warm, and life was flourishing. However, within 100 years after the event, virtually all life on Earth was destroyed. I would submit that humans could do well dealing with the 10 million year later conditions, but we should avoid, if at all possible, those conditions occurring during, and for a few hundred years after, the P/T catastrophe. The time bomb is the release of methane from the seafloor, and the annihilation of most life on Earth is the event we must prevent at all costs. Ultimately, we can deal with a warmer Earth; we just can't deal with it warming catastrophically.

Protecting Earth's Oceans

No human activity or endeavor can avert calamity if the ecological balance in Earth's oceans is destroyed. We cannot remove enough CO_2 to replace the photosynthesis currently taking place in the oceans regardless of what we plant on land and/or remove through chemical reactions. Since most agree that CO_2 levels in the atmosphere determine Earth's average temperature balance, the ocean's role in maintaining this balance is pivotal. Unfortunately, we have seen the pH of Earth's oceans change from 8.2 to 8.1 in the past 20 years or so (Wingenter et al 2007). Current estimates are that, unless things change drastically, the pH of the oceans will be 7.8 within the next 40 to 60 years.

We know that if we were to cease all activities leading to the acidification of Earth's oceans, the ocean pH would continue to lower for 20 years before equilibrium would exist. We also know that the bulk of the photosynthesizing and carbonate-producing organisms in the oceans are pH-sensitive, and many do not survive well in lower pH environments. By lowering pH to the levels predicted by the end of this century, virtually all marine photosynthesizers could be destroyed.

Taken together, these numbers give us between 20 and 40 years to completely cease contributing to the ocean's acidity, or we will have pushed the ocean pH beyond the survival boundaries of many marine organisms and thus set off a chain reaction we could never hope to mitigate.

Most of the pH changes in Earth's oceans are the result of acid rain – rain carrying carbon dioxide and other chemicals into the ocean from the atmosphere. In order to affect the pH of the oceans in a meaningful way, we would have to, within the next few years, eliminate our CO_2 output levels from fossil sources almost entirely. We would also have to curtail CO_2 production from modern sources until the system stabilizes. In order to accomplish these goals in the time allotted, we must convert our energy supply to sources not requiring fossil fuels and mostly not requiring carbon-based reactions.

Methane Clathrate Collapse

Were it not for this destructive timebomb ticking away under Earth's oceans, humans could probably allow nature to take its course without danger. However, virtually no life on Earth will survive if the frozen methane ice melts rapidly, and its melting is inevitable. The only real hope for human life to survive on Earth is to find a way to convert the existing methane clathrate and the methane gas trapped beneath its frozen layers, into an inert substance. This is a massive job. We have been burning fossil fuels at a tremendous rate for generations, and yet we have not used an amount equivalent to 10% of what is trapped in and beneath the clathrates on the ocean floor.

We have only just begun to explore our options for dealing with this substance. By the time we would be able to deal with this problem in a significant manner, it could be already released. To compound the problem, these deposits lie below the continental shelves that provide most of the habitat for marine fisheries. These fragile ecosystems would obviously be disrupted by either human mining activities or natural methane release. Unless we can come up with some practical solutions, these ecosystems are dead in a few generations either way.

Preventing a methane clathrate collapse is impossible. It is a natural process that has restored balance to the Earth's system several times in the past

– on the average of once every couple hundred million years. Unfortunately, we are living at the far end of that time cycle, and we may see the next correction event firsthand. Worse, we are likely providing the triggering mechanism that was supplied by volcanoes in the past – massive releases of CO_2.

The answer to this problem, as I see it, is fairly straightforward. We need to do three things to eventually bring the system back into balance without catastrophic destruction. First, we need to MASSIVELY reduce the amount of CO_2 in the atmosphere so that, 20 years from now, the pH of the oceans can once again stabilize. Second, we need to sequester carbon in stable forms so that we are no longer providing more carbon energy entrapment, and global temperatures can stabilize. And third, we need to convert the existing methane clathrate and associated trapped methane gas into stable compounds that won't release catastrophically. As an additional constraint, we need to do the above 3 things before the Earth experiences its next major volcanic eruption or solar event that would raise global temperatures and trigger the disaster regardless of our activities as humans.

Only by taking serious and possibly painful steps proactively can we hope to stave off the otherwise inevitable cataclysm. The following is a recap of the 3 steps mentioned above with some practical methods of dealing with the issues. Unfortunately, this is a global problem. There are no simple "fixes" to be applied. This is where we, as humans, make the ultimate decisions to either modify the Earth's systems through massive, worldwide alterations to our way of life, or we do as species have always done – live our lives as we wish, then go extinct when massive climate change happens.

Step 1: Massively Reduce CO2 in the Atmosphere and Oceans

As discussed earlier, the reduction of CO_2 has two aspects: the total amount of CO_2 being added to the atmosphere and the amount being added from fossil sources. As an example of the difference, one can compare money spent at a local business. If you spend money at a locally-owned business, that business owner and employees take their share and spend a significant portion in the community as well. This process continues to the point where the average dollar turns over seven times in the local community. However, a dollar spent at a national chain store only returns 15% to the local community through employee spending, etc. The bulk of the money spent leaves the community immediately.

Like a dollar spent at a local store, carbon dioxide vented to the atmosphere keeps recycling until some reaction removes it from the system permanently – sequesters it. These reactions occur when carbon is used to build shells, coral reefs, and other permanent structures, and also through the formation of coal. Unfortunately for us, current CO_2 removal also occurs through the formation of methane clathrate. In the past, the bulk of the Earth's carbon was removed through the formation of vast, stable coal deposits instead of the current formation of metastable methane ice. Ideally, to cure our current climate problem, we would find a solution where, like spending money in a chain store, the bulk of the carbon would leave the system permanently and not add to the methane problem.

Most of the CO_2 already added to the system from the past few hundred years of burning fossil fuels is still cycling through the biosphere. Current research concludes that this process will have a tremendous effect on the Earth's normal warm/cold cycles:

Carbon cycle models indicate that 25% of CO_2 from fossil fuel combustion will remain in the atmosphere for thousands of years, and 7% will remain beyond one hundred thousand years (Archer, 2005). We predict that a carbon release from fossil fuels or methane hydrate deposits of 5000 Gton C could prevent glaciation for the next 500,000 years, until after not one but two 400 kyr cycle eccentricity minima. The duration and intensity of the projected interglacial period are longer than have been seen in the last 2.6 million years. (Archer and Ganopolski, 2005)

We have already severely disrupted the carbon balance in our biosphere. Luckily for us, the Earth's systems have the ability to, at least temporarily, balance out such geologically rapid events. Unfortunately, the Earth tends to exhibit massive, long-term, life-destroying events when this balancing ability is exceeded. To limit CO_2 effectively, we would need to eliminate all coal burning within 20 years by:

A. Making maximum hydroelectric production from all major dam outlets mandatory, except for small flows related to maintaining migration patterns (fish ladders, etc.), and mandating the use of hydroelectricity as primary rather than as backup for the grid.

B. Creating additional nuclear electricity sources. However, as shown in Chernobyl and Japan, these facilities must be built in geologically stable areas and with appropriate redundant safeguards.

Do we do as species have always done? Live our lives as we wish, then go extinct when massive climate change happens?

There is an old saying: "It isn't what you do, it's how you do it." While nuclear energy production has inherent problems, those problems can be resolved through proper facility design, maintenance, and operation. This has been proven via the hundreds of nuclear power plants worldwide that have operated flawlessly for decades. We have the technology; we just need to implement it properly.

C. Investing in geothermal electrical production. This is, literally, an untapped resource. It is a commercially viable option wherever heated magma approaches the Earth's surface. Hot springs occur in many places throughout the world, and these usually indicate the presence of the appropriate heat within drilling distance of the surface. There has been no environmental downside identified with this method of electrical production. However, this technology has languished because funding has traditionally gone to support wind and solar projects instead.

D. Converting existing coal-fired plants. Currently operating coal-fired plants could easily be converted to burn fuel derived from modern plants rather than fossil fuel sources. Trees currently being allowed to burn naturally in forests – especially those already killed by insect infestation or other causes – could replace coal as the heat source for many of the existing generating plants. Remember – carbon from modern sources is not the main problem; sequestered carbon is. Modern plant carbon used in this manner is much less harmful, especially if the CO_2 produced is then itself captured, biologically altered, and sequestered as charcoal. In that manner, such former coal-fired electrical generating plants could become one of our major resources for curing the problem they once caused.

Hydrogen can be cheaply and easily produced, and hydrogen could be used to fire some of the other generating plants currently burning coal. Hydrogen is produced with relative ease through a number of processes. One of the most promising, from the perspective of climate benefit, is that methane may be stripped of its hydrogen atoms, leaving pure carbon behind as waste. This process is an ideal candidate for converting existing methane clathrate deposits into permanently sequestered carbon. As an added benefit to the reduction of methane clathrate in this manner, most internal combustion engines, such as those found in automobiles, can be converted to burn hydrogen with very little modification.

The solution does not have to be, "one size fits all." Power generating plants can be converted based on what makes the most sense based on their locations and designs. The important point is that they all can be, and should be, converted.

E. Use methane from clathrates. This compound has to be converted into a less volatile form if its catastrophic release is to be averted. There is enough energy stored in clathrates to supply the Earth's energy needs for hundreds of years. The only caveat is that both mining and burning processes for this substance need to be done in specific ways.

Mining needs to be done so as to not precipitate undersea events such as landslides and gas escapes. We do not yet have this technology, and we need to develop it rapidly. Major undersea structures exist that show what happens when methane clathrate releases incorrectly—large-scale undersea landslides can occur. While it is imperative that we remove this substance, we must do so in a way that does not cause other catastrophic events.

On the other end of the clathrate use processes, any CO_2 produced when the methane is burned needs to be sequestered. I would suggest that a viable method for sequestering would be to use biologic scrubbers that produce algae, dry the algae, turn it into charcoal, and bury it.

A better alternative, as mentioned earlier, may be to simply convert the methane into hydrogen gas and pure carbon as the primary process. Hydrogen gas burns by uniting with oxygen to produce water as the only reaction product. Imagine heating our homes, driving our cars, and powering our businesses, and seeing only water being vented out the smokestacks and automobile exhausts! The only downside to this method would be the eventual increase in the amount of water on the planet. To balance, however, we would also be able to produce significant quantities of pure fresh water wherever hydrogen fuels were used in quantity.

Within the realm of current technology, pure carbon and charcoal are the only viable solution for sequestering carbon cheaply, efficiently, and without further disrupting the environment. However, as time goes by, other options may become available.

Only through these actions can we, as Earth's inhabitants, significantly limit our current impact on our planet's carbon dioxide volume. We are already too late to stop a tremendous rise in Earth's global temperatures. However, we may not be too late to stop the catastrophic temperature spike that will destroy life as we know it.

Geothermal electrical production is, literally, an untapped resource.

Step 2: Permanent Sequestration of Carbon

In order to maintain or reduce the current level of carbon in the biosphere, we must sequester at least as many carbon molecules as are being released. Also, we must always bear in mind the potential energy entrapment potential for any given molecule. For instance, a methane molecule can trap 20 times the energy compared to a CO_2 molecule, and when it burns, it produces a CO_2 molecule in a ratio of 1:1. Therefore, simple arithmetic dictates that it is 20 times more important to remove methane from the atmosphere as it is to remove carbon dioxide. Carbon dioxide must be our primary concern at present due to its sheer volume in the biosphere. However, a relatively small methane release could do as much damage to Earth's climate balance as a CO_2 release 20 times as large.

The first part of planning for carbon sequestration needs to be committing to realistic expectations and desired results. The only significant budget for carbon sequestration research has been devoted to studying the possibility of injecting carbon dioxide gas deep underground. Like planting trees, this process, once perfected, may provide short-term benefits. However, over the long term, the volumes and pressures such sequestration would require would be impossible to maintain. Sooner or later, the container would weaken and the gas would escape. There is no way known to man to seal massive areas underground and permanently maintain that seal through earthquakes, volcanic events, or other ground-shifting activities. Such sequestration methods may serve political purposes or the needs of other industries, but there is no guarantee that such sequestration would be viable in the long term.

The main method of carbon sequestration for the Earth at present is the continued formation of methane clathrate, and this is an unacceptable solution. Ultimately the survival of the human race and most other life on Earth depends on most of Earth's methane clathrate being converted to a more stable sequestered form. This can be done in the form of carbonate rocks, charcoal, polymerized hydrocarbons, and silicon carbide, to name a few options. Most of these were discussed earlier in this work. To recap the points made previously, carbonates are a less desirable option due to their required oxygen component. Silicon carbide is the ideal solution, but it is produced at too high an energy cost. Polymerized hydrocarbons vary in stability, and it is likely impractical to deliberately produce one simply for the purpose of carbon sequestration. That leaves charcoal as the most viable option for current carbon sequestration.

One possibility that should be mentioned is a multi-stage approach. Landfills containing well-compacted, oxygen-deprived, buried polymerized hydrocarbons may be a good short- and mid-term solution. The compaction and burial of automotive tires could be a solution to two different problems. Plastics and cellulose could also be used for carbon sequestration in the short term. Such burials tend to produce methane, but the methane could be captured and recycled.

The good news is that the end product of their breakdown would resemble current natural deposits of chain carbons – asphalt, tar, oils, coals, and such. It is possible that some of Earth's carbon balance could be restored temporarily through the production of vast landfills full of shredded tires, certain plastics, and paper (cellulose). However, these deposits would have to be carefully designed and monitored so that any breakdown products could also be captured and sequestered. Current landfills for human garbage are not properly designed or maintained for this purpose.

The only easily accomplished, reliable, permanent, and inexpensive sequestration for Earth's carbon is in the form of coal. We can produce charcoal, which is, in effect, modern coal, simply by baking cellulose materials in an oxygen-reduced environment. Heat for this baking could be provided by intermittent energy sources, such as solar power, that are less suitable choices for electrical generation due to their intermittent natures. However, when one considers the number of tons of coal currently being burned each day, the task of sequestering equal volumes is daunting. The sequestration process must, therefore, include all of the following:

A. Short-term solutions such as planting trees.

B. Short- to mid-range solutions such as sequestering human by products containing cellulose and long chain hydrocarbons.

C. Long-term sequestration through coal and carbonate production.

It took the Earth hundreds of millions of years to sequester carbon to the levels seen a few hundred years ago, but it has only taken humans a few hundred years to introduce a significant portion back into the biosphere. Carbon is easy to release, but difficult to sequester. It will take a determined, multi-faceted approach for humans to re-sequester the carbon they have recently released. It will take even more effort, however, to perform that sequestration PLUS sequester the additional carbon being re-introduced naturally through the melting of clathrates, the breakdown of carbonate rocks through acid rain and other factors, and through volcanic events. However, this must be done, and the balance maintained, until we can defuse Earth's temperature time bomb: methane clathrate.

Step 3: Remove the Methane Clathrate Threat

Ultimately, neither human prosperity nor human survival is the issue raised in this volume. The real issue is the potential extinction of virtually all life on Earth in the relatively near future. How we, deal with this problem will determine whether most other life forms currently residing on this planet still exist on Earth in a few hundred years as well as our own species.

It was only a couple of decades ago that we became aware that this substance existed in quantity, and it was only a few years ago any of us realized the deadly nature of this substance. Between the justifiable caution of scientists wishing to be sure of their data before sounding an alarm, the even-more-justifiable desire not to cause undue panic in the world's human population, and the not-so-justifiable worries of disrupting current political and economic balances, the public is largely unaware of the scope of this problem.

What Can You Do?

Unfortunately, this is a global problem. As individuals working alone, we can do little to stave off this catastrophic event. However, the Earth's collective governments do have the power to make the changes necessary for our survival. We need to work together to insist that those representing us make the timely and necessary decisions to force the changes in our collective energy supplies. This is our first and foremost mission, and it requires each and every one of us to apply the necessary pressure to force the change.

Individually, we also need to make lifestyle changes. If we, as individuals, pay more attention to the energy balance in our daily lives we can begin to

Figure 31. We must all be part of a global effort to deal with the situation.

see where improvements may be made. Purchases made based on true climate-friendly characteristics will drive producers to make climate-friendly decisions. A little inconvenience on our parts individually can exert a tremendous economic and political pressure collectively. An electric car is a wonderful thing if it's plugged into a solar panel for recharging rather than the grid. On the other hand, ethanol uses more fossil fuel to create than it replaces, on average. Heating a home by burning wood (properly catalyzed), solar, or geothermal heat pump can save tons of sequestered fossil fuels.

Most importantly, we need to make sure the entire world understands that the problem we face can only be solved by increasing the amount of buried carbon, not decreasing it. Any process that takes carbon out of the system, if even for a few hundred years, is desirable. Conversely, any process that adds carbon to the system from sequestered sources is a problem. The exception is the burning of methane clathrate, capturing the exhaust CO_2, converting it to pure carbon, and reburying it. We need to burn gigatons of clathrate and convert it as rapidly as possible. We need to defuse the bomb.

The Good News

The majority of this book has been focused on a recently discovered, potentially catastrophic problem. However, the fossil record also shows that life can survive and prosper even if current climate conditions are dramatically altered. We only have to extend the period of time during which methane clathrate is released and prevent the oceans from being sterilized. We can survive as a species, and quite possibly without significant disruption in our

Figure. 32. Carbon is easy to release, but difficult to sequester.

current ways of life, if global temperatures increase – just not if they increase catastrophically. Just like when a middle-age person is told that blood sugar or heart conditions require minor lifestyle changes in order to maintain quality-of-life, so have we now been told that we have to make lifestyle changes in our species' energy consumption in order to survive and stay healthy.

Imagine a world where oceans are warm, and climate is subtropical virtually from equator to the poles. Imagine oxygen levels that allow easy breathing at 20,000 feet. Fires may start easier and burn faster, but greater humidity in the air makes them easier to put out. Plants grow profusely because of high levels of carbon dioxide in the atmosphere as well as oxygen. This is the world in which the dinosaurs lived. This is the world of the late Cretaceous Period. Our world is stable under those conditions, and I believe we have the ability to engineer our environment, over time, to achieve that stability.

The methane clathrate threat is real. However, we don't have to completely eliminate that threat. We only have to arrange it so that the majority of ocean life survives the period of instability. We don't need to convert all, or even most, of the CO_2 into sequestered carbon. We just need to make sure that its increased presence doesn't trigger a clathrate collapse. And, finally, we need to stop worrying about events that raise and lower sea levels a few meters. This is normal variation in Earth's cyclic nature. Live with it. Visit the beach, don't live there.

We cannot create a truly stable Earth. The Earth is too large and too complex. However, we can tweak the existing global system so that the changes don't destroy the entire ecosystem as has happened in the past. Scientists are searching the skies, hoping to chart the next big rock from space that will hit us. They know that a small push in the right direction, if done early enough, will avert disaster. The same is true with our current climate. We would never be able to stop the clathrate collapse once it started. However, relatively small pushes now, ones that are within our capabilities, will prevent that catastrophe later.

However, the warning is coming late – at a time when the Earth is already exhibiting climate changes that can be reasonably linked to human activities already performed. Due to the size of global human infrastructure, even a warning heeded would take years to see the appropriate modifications performed. With the stakes so high – survival itself – and the window of opportunity so short – maybe as little as 20 years – I believe that we cannot afford to risk complacency or procrastination. We are in a race between our activities and the point in time when we are no longer capable of doing enough to stave off the catastrophe. And the wager on this race is life itself. We can win this wager.

We must.

We will.

The wager on this race is life itself. We can win this wager. We must. We will.

F.A.Q.: Answers to Frequently Asked Questions

For those of you who are used to going directly to the FAQ's page of a website or those who have to flip to the back of the book to see how it ends, this section is for you. I presented detailed explanations of why these answers are given in previous chapters, but perhaps not as succinct or focused as presented here. Many of these answers differ from the current popular understanding of the situation, and I encourage my readers to examine the data for themselves before passing judgment on the accuracy of the answers given below.

Is Global Climate Change ("Warming") "real"?

Yes.

Is the current global climate change "human caused?"

Partially, but not entirely. It is largely a natural process.

Is global warming inevitable?

Yes. The Earth will warm. The only questions are "how much?" and "how fast?"

So, what's the problem?

If the temperature rises naturally, as it has in the past, it will do so catastrophically. If this happens, most, if not all, humans will perish in the event and most other life on Earth will also perish. We must do all in our power to see that this change happens gradually if we want the human species to survive.

Is the current global warming event a serious threat to human existence?

Yes.

Will the Earth continue to get warmer each year, and will all areas on Earth be equally affected?

No.

Are the current "green" policies helpful at all?

Probably not.

If we observe a leveling off or a cooling in climate over the next few years, is the problem solved?

No.

Is the potential climate change we are facing now greater or lesser than that faced by the dinosaurs at the time of their extinction?

Greater.

Is there another time in Earth's history that matches climate and events almost identically with the present?

Yes, the Permian-Triassic boundary, when over 90% of all life on Earth was wiped out.

What human activities are contributing to the warming event?

Not as many as are currently reported. The list can actually be limited to any activity that takes buried carbon and puts it back into the biosphere.

Can we create worse problems for human life on Earth by trying to fix the current problems without proper forethought and planning, and especially by not utilizing paleontologic data?

Yes.

Can we prevent a catastrophic global warming event if we act now and in the necessary manner?

Probably.

Can we still do the same if we wait 30 years?

Possibly not. The system is so massive and complex that small changes now will provide a better end result than even large changes later on. The problem is so complex that we cannot determine, exactly, the "point of no return."

What is the main cause of the demise of the Dinosaurs?

Rapid climate change.

Throughout the 4.5 billion years of Earth's geologic time, what single factor has affected the amplitude of global temperature changes the most?

The amount of carbon, relative to the other elements, in the biosphere. It wasn't until most carbon was taken out of the system that Earth's climate became habitable for complex life. For comparison, think of Venus. Venus could be a habitable planet were it not for its high atmospheric CO_2 levels and resultant runaway greenhouse effect.

How can you tell which carbon is a problem and which is not?

Carbon that does not interact with the biosphere – that trapped in coal, carbonate rocks, etc. – is not a problem. All other carbon contributes to the problem, including that trapped in plants, gases trapped undersea (they slowly seep out), and even the gases belched out of volcanoes.

What forms of carbon are of particular concern?

Carbon in plants and animals, carbon in gases (methane, carbon dioxide, carbon monoxide, etc), carbon in short chain, unstable hydrocarbons, carbon trapped in unstable environments (such as the methane clathrate on the ocean floor) and carbon that humans are converting from stable to unstable forms.

What, exactly, is the problem with global warming proceeding naturally?

The Earth could potentially lose half its free oxygen, sterilize the oceans, and warm at least 10 degrees Celsius (nearly 20 degrees Fahrenheit), all likely within a time span of less than 20 years from start to finish.

What does carbon have to do with the oxygen in our atmosphere?

Originally, there was no free oxygen. Carbon dioxide was broken down by photosynthetic plants. These plants produced carbon compounds for their structures from water and carbon dioxide. Oxygen was given off as a by-product of these reactions. Thus, if we were to burn all the free carbon – coal, plants, methane, etc., on the Earth, we would have no oxygen left in the atmosphere.

What event could possibly cause such radical changes to our biosphere?

The release of over 50% of the Earth's trapped hydrocarbons through the melting of the Earth's methane ice (clathrate) currently on the ocean floor and under the tundra.

What is methane hydrate/clathrate/ice?

It is a special combination of frozen water and methane that is formed under pressure. It is present on virtually all continental shelves on the ocean floor, and each liter of methane ice traps an average of over 160 liters of methane gas at standard atmospheric pressure.

Which greenhouse gas has the potential for doing the most damage?

Methane. Also known as natural gas, this substance is poisonous to most life; it has over 20 times the greenhouse effect of carbon dioxide, and after it does its damage in the system it oxidizes (burns) to produce lot of heat. In the process, it removes 2 oxygen atoms from the atmosphere for each molecule of methane oxidized, and it forms carbon dioxide molecule-for-molecule. Not only is it a potent greenhouse gas; it also exists in quantities to as sure total climate disruption on Earth should it be released.

Why is this crisis so catastrophic?

The release of methane (natural gas) in such quantities as clathrate breakdown would cause has the potential for great disruption. First, it poisons the oceans as it bubbles up from the bottom. Once it reaches the atmosphere it forms poison gas clouds. The buildup of methane past a certain level in the atmosphere would cause regional flash fires. Methane is over 20 times more effective as a greenhouse gas than carbon dioxide; and when it breaks down, it produces carbon dioxide molecule-for molecule. Such changes in our atmosphere allow more of the Sun's energy to be trapped, causing the Earth's average temperature to be much warmer (over 10 degrees Celsius, 18 degrees Fahrenheit) for millions of years following such an event.

What will cause this release?

The oceans are currently removing most of the CO2 in our atmosphere. However, several factors mentioned in this book, including changes in ocean temperatures or pH levels or the loss of polar ice caps, may cause the oceans to lose their ability to remove carbon dioxide from the system. Once that happens, the CO2 levels in the atmosphere will spike. The CO2 spike will, in turn, cause rapid warming, triggering the release of methane from the ocean floor.

How much carbon was in the biosphere 20,000 years ago during the last ice age advance?

It was near, or at, an all-time low.

What single human activity is responsible for returning the amount of carbon in the biosphere to pre-ice age levels and causing over 60% of the increase in greenhouse gas emissions?

Coal-fired heating and electricity generation.

Why is the burning of coal a greenhouse gas problem?

The chemical equation for generating power from coal is C (carbon – coal is virtually pure carbon) plus O2 (oxygen) yields CO2 (carbon dioxide). You can't burn coal to produce power without releasing an equivalent amount of carbon dioxide and consuming oxygen.

How can we tell when we are on the brink of disaster?

The oceans remove most of the planet's greenhouse gases. When the oceans start faltering in this process we are in real, imminent trouble. Without their proper functioning it is a matter of a few years at most before temperatures begin to rise catastrophically.

What biologic cause would quickly reduce the ability for the oceans to remove carbon dioxide from the atmosphere?

Too much of a good thing. The ocean's green plants flourish on carbon dioxide, but they are very pH sensitive. Unfortunately, too much CO2 is not a good thing for many of the ocean's organisms. Dissolving of atmospheric CO2 into the water and subsequent formation of carbonic acid is even now lowering the pH of the Earth's oceans. Once the pH of the oceans reaches a certain point, the oceans' plants and microorganisms will start dying off. Loss of these organisms will eventually eliminate 70% of Earth's ability to remove CO2, and its buildup in Earth's atmosphere will increase exponentially.

What other event could cause the methane release?

Lack of cold, fresh water circulating at the bottom of the oceans. Should this happen, the methane ice would melt, releasing the methane gas. Such currents are produced from polar melt flow. If we lose the polar ice caps, we lose this flow, and the methane would then be released.

What are the most important steps in preventing a runaway global warming event that we can't survive?

Stop burning coal. Use alternative electricity generation methods (wind, solar, hydrogen, geothermal, hydro, nuclear, etc.) instead. Use methane clathrate (while properly sequestering the carbon by-product) for much of our energy needs so that we can remove it from Earth's system. Sequester pure carbon (charcoal) and complex hydrocarbons such as inert plastics and tires. And, most importantly, fund further TRUE global climate research including paleontology and Earth history, and also fund proper, coordinated global climate monitoring and adjustment.

We must not cause the next ice age by trying to mitigate global warming. We must all realize that the Earth needs to warm significantly just to bring it back to its equilibrium point, but we need this to happen gradually rather than as the result of a methane catastrophe.

Global warming is a natural and necessary event for the Earth itself. However, mass extinction is the price if this event is allowed to proceed naturally. We probably cannot stop the event, but we can modify the event so that we can possibly survive it. But, in order to modify it in time, we must act soon.

George Santayana (1905) once said, "Those who cannot remember the past are condemned to repeat it." I would go a step farther and say that those who cannot learn from the Earth's past are going to end up like the dinosaurs. A rapid climate change, a little dust in the atmosphere, and the loss of the preferred edible plants to a very robust weed (grass), and the dinosaurs found themselves unable to adapt to the new conditions rapidly enough to survive. For the current level of human civilization, relatively minor (in terms of Earth's past history) environmental changes would cause catastrophic mortality due to economic and infrastructure collapse. A major environmental change, such as the ones that occurred 670 million years ago, 444 million years ago, 251 million years ago, or 65 million years ago would be the end of human civilization, and it could possibly spell the end of the human species.

Unlike the dinosaurs, however, we have been warned. We have the understanding and capability to prevent this catastrophe. In a way, the dinosaurs gave their collective lives so that we could learn from their past experiences. It would be a shame if such a warning went unheeded.

Unlike the dinosaurs, we have been warned. We have the understanding and capability to prevent this catastrophe.

Acknowledgments:

This work is a synthesis of ideas and data from literally hundreds of sources. Without the efforts of these researchers, such a synthesis would not have been possible. I thus offer my heartfelt gratitude to all those who have fought the elements, politics, and economics to gather the data and make the inferences. These researchers are the true heroes of this story. And, while I have recognized a number of these individuals through citations and references, many more have contributed without recognition.

I would also like to thank the several people over the years who have offered their assistance to this project, including but not limited to (in alphabetical order) Michael Berglund, Cory Coverdell, Sue Frary, Lotus Grenier, Erin King, Matt Trexler, and Laurie Trexler. I cannot adequately express my gratitude for their diligence in reading manuscripts, assisting with research, and formatting this work. Without them, this book would still be an unpublished file on my computer.

Most of all, I wish to thank my family and friends. You have put up with my one-track mind, my frustrations, and my unavailability both at home and at work for the sake of finishing this project. You will, by this time, have received the first hardcover copies of this book. I hope you will treasure it for its intrinsic value. However, if you burn it as token recompense for the past few years of suffering, I will understand! If we are correct in our interpretations, the heat of that fire, at least, will not contribute to the carbon imbalance!

—David Trexler

References

Archer, D. (2005). Fate of fossil fuel CO2 in geologic time. Journal of Geophysical Research, 110 (C09S05), 6 pp. DOI: 10.1029/2004JC002625

Archer, D., & Ganopolski, A. (2005). A movable trigger: Fossil fuel CO2 and the onset of the next glaciation. Geochemistry Geophysics Geosystems 6 (Q05003), 7 pp. DOI: 10.1029/2004GC000891

Beerling, D. J., Harfoot, M., Lomax, B., & Pyle, J. A., (2007). The stability of the stratospheric ozone layer during the End-Permian eruption of the Siberian Traps. Philosophical Transactions of the Royal Society A, 365, 1843-1866. doi: 10.1098/rsta.2007.2046

Berger, A., (1988). Milankovitch theory and climate. Reviews of Geophysics, 26 (4), 624-657.

Berger, A., Loutre, M. F., & Crucifix, M., (2003). The Earth's climate in the next hundred thousand years (100 kyr). Surveys in Geophysics, 24, 117-138.

Berner, R. A., (2002). Examination of hypotheses for the Permo–Triassic Boundary extinction by carbon cycle modeling. Proceedings, National Academy of Science, 99 (7), 4172–4177.

Berner, R.A., (2006). GEOCARBSULF: A combined model for Phanerozoic atmospheric O2 and CO2. Geochimica et Cosmochimica Acta 70, 5653-5664.

Berner, R. A., & Kothavala, Z., (2001). Geocarb III: A revised model of atmospheric CO2 over phanerozoic time. American Journal of Science, 301, 182-204

Boggs, S. Jr., (2001). Principles of Sedimentology and Stratigraphy, 3rd ed. Upper Saddle River, NJ: Prentice Hall.

Bouvier, A., Blichert-Toft, J., Moynier, F., Vervoort, J. D., & Albarède, F., (2007). Pb-Pb dating constraints on the accretion and cooling history of chondrites. Geochimica et Cosmochimica Acta, 71, 1583-1604.

Bottjer, D. J., Clapham, M. E., Fraiser, M. L., & Powers, C. M., (2008). Understanding mechanisms for the End-Permian mass extinction and the protracted Early Triassic aftermath and recovery. GSA Today, 18 (9), 4-10.

Broecker, W. S., (1997). Thermohaline circulation, the achilles heel of our climate system: will man-made CO2 upset the current balance? Science New Series, 278 (5343), 1582-1588.

Buffett, B., & Archer, D., (2004). Global inventory of methane clathrate: Sensitivity to changes in the deep ocean. Earth and Planetary Science Letters 227, 185-199.

CCSP Research Highlight, (2006). "Methane as a greenhouse gas" Retrieved from http://www.climatescience.gov/infosheets/highlight1/default.htm

Chatti, I., Delahaye, A., Fournaison, L,. & Petitet, J. P., (2005). Benefits and drawbacks of clathrate hydrates: A review of their areas of interest. Energy Conversion and Management 46,1333-1343.

Cohen, J. E., (2003). Human population: the next half century. Science New Series, 302 (5648), 1172-1175.

Cole-Dai, J., Ferris, D., Lanciki, A. L., Savarino, J., Baroni M., & Thiemens, M. H., (2009). Cold decade (AD 1810-1819) caused by Tambora (1815) and another (1809) stratospheric volcanic eruption. Geophysical Research Letters 36, L22703.

Coverdell, C., (July, 2010) Two Medicine Dinosaur Center. personal communication.

Dorritie, D., (2007). Killer in our midst: methane catastrophes in Earth's Past . . . and near future? Retrieved from www.killerinourmidst.com.

Feddema, J. J., Oleson, K. W., Bonan, G. B., Mearns, L. O., Buja, L. E., Meehl, G. A., & Washington, W. M., (2005). The importance of land-cover change in simulating future climates. Science 310, 1674-1678. doi: 10.1126/science.1118160

Fortin, D., & Langley, S., (2005). Formation and occurrence of biogenic iron-rich minerals. Earth Science Reviews 72, 1-19. doi:10.1016/j.earscirev.2005.03.002

Friis-Christensen, E., & Lassen, K., (1991). Length of the solar cycle: An indicator of solar activity closely associated with climate. Science 254, 698-700.

Gabbard, A., (1993). Coal combustion: Nuclear resource or danger. Oakridge National Laboratory Review (ORNL) 26. Retrieved from http://www.ornl.gov/info/ornlreview/rev26-34/text/colmain.html
Gas escaping from ocean floor may drive global warming. Office of Public Affairs, UC Santa Barbara. July 19, 2006. Retrieved from http://www.ia.ucsb.edu/pa/display.aspx?pkey=1482

Glasby, G. P., (2006). Drastic reductions in utilizable fossil fuel reserves: An environmental imperative. Environment, Development and Sustainability 8,197-215. doi: 10.1007/s10668-005-5753-4

Glatzmaier, G.A., Coe, R.S., Hongre, L., & Roberts, P.H., (1999). The role of the Earth's mantle in controlling the frequency of geomagnetic reversals. Nature 401, 885-890.

Glatzmaier, G.A. & Roberts, P. H., (1995). A three-dimensional self-consistent computer simulation of a geomagnetic field reversal. Nature 377: 203-209.

Gradstein, F.M., Ogg, J.G., & Smith, A. G. (Eds.). (2004). A Geologic Time Scale 2004. Cambridge: University Press.

Hansen, J., & Sato, M., (2004). Greenhouse gas growth rates. Proceedings, National Academy of Science 101 (46), 16109-16114

Hetherington, R., & Reid, R. G. B., (2010). The climate connection: climate change and modern human evolution. Cambridge: University Press.

Hill, T. M., Kennett, J. P., Valentine, D. L., Yang, Z., Reddy, C. M., Nelson, R. K., . . . Beaufort, L., (2006). Climatically driven emissions of hydrocarbons from marine sediments during deglaciation. Proceedings of the National Academy of Sciences of the United States of America 103 (37), 13570-13574.

Hoffman, K. A., (1995). How are geomagnetic reversals related to field intensity? Eos 76, 289.

Honjo, S., (1997, December 1). Marine snow and fecal pellets: the spring rain and food to the abyss. Oceanus. Retrieved from http://www.whoi.edu/oceanus/viewArticle.do?id=2387

Hopkin, M., (2007). Carbon sinks threatened by increasing ozone. Nature 448,396-397. doi:10.1038/448396b

Jacobson, M. Z., (2004). The short-term cooling but long-term global warming due to bio mass burning. Journal of Climate 17, 2909-2926.

Jin, Z,, Charlock, T. P., Smith, W. L. Jr., & Rutledge, K., (2004). A parameterization of ocean surface albedo. Geophysical Research Letters 31 (L22301), 1-4. doi:10.1029/2004GL021180

Kasting, J. F., & Howard, M. T., (2006). Atmospheric composition and climate on the early Earth. Philosophical Transactions of the Royal Society B 361 1733-1742. doi: 10.1.098/rstb.2006.1902

Katz, M., Cramer, B. S., Mountain, G. S., Katz, S., & Miller, K. G., (2001). Uncorking the bottle: What triggered the Paleocene/Eocene thermal maximum methane release? Paleoceanography 16 (6), 549-562.

Kearsey, T., Twitchett, R. J., Price, G. D., & Grimes, S. T., (2009). Isotope excursions and paleotemperature estimates from the Permian/Triassic boundary in the Southern Alps (Italy) Paleogeography, Paleoclimatology, Paleoecology 279, 29-40.

Kiehl, J. T. & Shields, C. A., (2005). Climate simulation of the latest Permian: Implications for mass extinction. Geology 33 (9), 757-760. doi: 10.1130/G21654.1

Kiehl, J. T. & Trenberth, K. E., (1997). Earth's annual global mean energy budget. Bulletin of the American Meteorological Society 78, 197-208.

Kious, W. J., & Tilling, R. I., (1996). This dynamic Earth: The story of plate tectonics. Washington, D.C: United States Geological Survey.

Knoll, A. H., Bambach, R. K., Payne, J. L., Pruss, S., & Fischer, W. W., (2007). Paleophysiology and the End-Permian mass extinction. Earth and Planetary Science Letters 256,295-313. doi:10.1016/j.epsl.2007.02.018

Kump, L. R., (2008). The rise of atmospheric oxygen. Nature 451, 277-278. doi:10.1038/nature06587

Kvenvolden, K. A., (1998). A primer on the geological occurrence of gas hydrates. In J. P. Henriet, & J. Mienert, eds., Gas Hydrates - Relevance to World Margin Stability and Climate Change (pp. 9-30). London: The Geological Society. doi: 10.1144/GSL.SP.1998.137.01.02

Laurance, W. F., (2010). Habitat destruction: death by a thousand cuts. In N. S. Sodhi & P. R. Ehrlich, (Eds.), Conservation Biology for All (pp. 73-87). Oxford: University Press.

Lisiecki, L. E., & Raymo, M. E., (2005). A Pliocene-Pleistocene stack of 57 globally distributed benthic d18o records. Paleoceanography 20 (PA1003), 17 pp. doi:10.1029/2004PA001071

McMurry, D., (June, 2010). Operations Manager, Fort Peck Reservoir, Army Corps of Engineers. Personal communication.

Miller, J. B., Gatti, L. B., d'Amelio, M. T. S., Crotwell, A. M., Dlugokenky, E. J., Baldwin, P., . . . Tans, P. P., (2007). Airborne measurements indicate large methane emissions from the eastern Amazon Basin. Geophysical Research Letters 34 (L10809), 5 pp. doi:10.1029/2006GL029213

Moussas, X., Polygiannakis, J. M., Preka-Papadema, P., & Exarhos, G., (2005). Solar cycles: A tutorial. Advances in Space Research 35, 725-738.

Prothero, D. R., (2009). Do impacts really cause mass extinctions? In J. Seckbach & M. Walsh (Eds.) From Fossils to Astrobiology (pp. 409-423). Dordrecht, Netherlands: Springer Science + Business Media B. V.

Pruss, S. B., & Bottjer, D. J., (2005). The reorganization of reef communities following the End-Permian mass extinction. Comptes Rendus Palevol 4 (6-7), 7553-568. doi:10.1016/j.crpv.2005.04.003

Rees, P. M., (2002). Land-plant diversity and the End-Permian mass extinction. Geology 30 (9), 827–830.

Retallack, G. J., (2005). Permian greenhouse crises. In S. G. Lucas & K. E. Zeigler, (Eds.), The Nonmarine Permian (pp. 256-269). Albuquerque NM: New Mexico Museum of Natural History and Science.

Retallack, G. J., Veevers, J. J., & Morante, R., (1996). Global coal gap between Permian-Triassic extinction and Middle Triassic recovery of peat-forming plants. GSA Bulletin 108 (2), 195-207. DOI: 10.1130/0016-7606(1996) 108<0195:GCGBPT>2.3.CO;2

Rieu, R., Allen, P. A., Plötze, M., & Pettke, T., (2007). Climatic cycles during a Neoproterozoic "snowball" glacial epoch. Geology 35 (4), 299-302. doi: 10.1130/G23400A.1

Running, Stephen, (January, 2008). Comments presented at a lecture at Choteau, Montana.

Ryberg, P. E., & Taylor, E. L., (2007). Silicified wood from the Permian and Triassic of Antarctica: Tree rings from polar paleolatitudes. In A. K. Cooper and C. R. Raymond et al., (Eds.), Antarctica: A Keystone in a Changing World – Online Proceedings of the 10th ISAES, Washington D. C.: USGS. Retrieved from http://pubs.usgs.gov/of/2007/1047/srp/srp080/of2007-1047srp080.pdf doi:10.3133/of2007-1047.srp080

Saier, M. H. Jr., (2010). Desertification and migration. Water, Air, & Soil Pollution 205 (1 supp.), 31-32. DOI: 10.1007/s11270-007-9429-6

Santayana, G., (1905). Reason in common sense. Life of Reason 1, 284.

Schmidt, G., (2004). Methane: A Scientific Journey from Obscurity to Scientific Stardom. NASA Research Feature: National Aeronautics and Space Administration, Goddard Institute for Space Studies. Retrieved from http://www.giss.nasa.gov/ research/features/200409_methane/

Scotese, Christopher R., (2003). Paleomap Project. Retrieved from www.scotese.com.

Sheehan, Peter M., (2001). The Late Ordovician mass extinction. Annual Reviews of Earth and Planetary Sciences 29, 331-364.

Shen, S. Z., Cao, C. Q., Henderson, C. M., Wang, X. D., Shi, G. R., Wang, Y., & Wang, W., (2006). End-Permian mass extinction pattern in the northern peri-gondwanan region. Palaeoworld 15, 3–30.

Sivakumar, M. V. K., 2007. Interactions between climate and desertification. Agricultural and Forest Meteorology 142, 143-155. doi:10.1016/j.agrformet.2006.03.025

Sudarshan, T. S., & Maximenko, S. I., (2006). Bulk growth of single crystal silicon carbide. Microelectronic Engineering 83,155-159.

Tans, P., (2009). An accounting of the observed increase in oceanic and atmospheric CO2 and an outlook for the future. Oceanography 22 (4), 26-35.

Tans, P., (2010). Trends in atmospheric carbon dioxide. Retrieved from www.esrl.noaa.gov/gmd/ccgg/trends/

Thompson, A. E., (2003). Fossil Fuel Facts. Global Strategies Forum, World Future Society. Retrieved from http://www.wfs.org/thompson03.htm

Thorpe, A., (2009). Enteric fermentation and ruminant eructation: the role (and control?) of methane in the climate change debate.
Climatic Change 93,407-431 doi:10.1007/s10584-008-9506-x

Tice, M. M., & Lowe, D. R., (2004). Photosynthetic microbial mats in the 3,416-Myr-old ocean. Nature 431, 549-552. doi:10.1038/nature02888

Trenberth, K. E., & Stepaniak, D. P., (2004). The flow of energy through the Earth's climate system. Royal Meteorological Society 130 (603): 2677-2701. doi: 10.1256/qj.o4.83

Uhl, D., Andrè Jasper, A., Abu Hamad, A. M. B., & Montenari, M., (2008). Permian and Triassic wildfires and atmospheric oxygen levels. Proceedings of the 1st WSEAS International Conference on Environmental and Geological Science and Engineering 1, 179-187.

Wacey, D., McLoughlin, N. & Brasier, M. 2009. The search for windows into the earliest history of life on Earth and Mars. In: J. Sechbach, (Ed.),Cellular Origin, Life in Extreme Habitats and Astrobiology, Volume 11. From fossils to Astrobiology. (pp.41-68). Berlin: Springer (V).

Wada, K., & Kokubo, E., (2006). High-resolution simulations of a moon-forming impact and postimpact evolution. The Astrophysical Journal (638), 1180-1186.

Weishampel, D. B., Dodson, P., & Osmólska, H., (Eds.) (1990). The Dinosauria (733 pp.). Berkeley, CA: University of California Press.

Wilde, S., Valley, J., Peck, W., & Graham, C., (2001). Evidence from detrital zircons for the existence of continental crust and oceans on the Earth 4.4 Gyr ago. Nature 409,175-178.

Wingenter, O. W., Haase, K. B., Zeigler, M., Blake, D. R., Rowland, F. S., Sive, . . . Riebesell U., (2007). Unexpected consequences of increasing CO_2 and ocean acidity on marine production of DMS and CH_2ClI: Potential climate impacts. Geophysical Research Letters 34 (L05710), 5 pp. doi: 10.1029/2006GL028139

Zhao, Z., Mao, XY., Chai, ZF., Yang, GC., Zhang, FC., & Yan, Z., (2009). Geochemical environmental changes and dinosaur extinction during the Cretaceous-Paleogene (K/T) transition in the Nanxiong Basin, South China: Evidence from dinosaur eggshells. Chinese Science Bulletin 54, (5), 806-815. DOI: 10.1007/s11434-008-0565-1

Websites

http://www.climate-leaders.org/climate-change-resources/climate-change/preventing-climate-change
http://www.scotese.com
http://www.esrl.noaa.gov/gmd/ccgg/trends/
http://www.wfs.org/thompson03.htm
http://www.giss.nasa.gov/research/features/200409_methane/
http://www.ia.ucsb.edu/pa/display.aspx?pkey=1482
http://www.nasa.gov/vision/earth/environment/ozone_resource_page.html

Agencies and Government Web Resources

Intergovernmental Panel on Climate Change: http://www.ipcc.ch/
NOAA Annual Greenhouse Gas Index: http://www.esrl.noaa.gov/gmd/aggi/
Kyoto Protocol: available online at http://unfccc.int/resource/docs/convkp/kpeng.html
US Energy Information Administration: http://www.eia.doe.gov/

Index

albedo, 12, 55-59
    as climate regulator, 19, 41
    rainforest loss, effect of, 12
    water vapor, effect of, 44

atmosphere,
    carbon dioxide (see carbon
    dioxide)
    in Earth's early history, 78
    energy entrapment (see energy
    entrapment)
    greenhouse gas (see greenhouse
    gas)
    Hadley Circulation, 30-32
    photochemical smog, 44
    of urban areas, 55
    water vapor (see water vapor)

banded iron, 78, 81

carbon. See also carbon dioxide. See also
coal
    emissions, potential reduction of,
    144-147, 158-168
    hydrocarbon compounds, 54-55
    life-formed molecules, 77-78
    oil, formation of, 126
    sequester potential, 122-131,
    142-144, 165-168

carbon dioxide
    carbonic acid, 35, 102
    deforestation, created by, 45
    in Earth's early history, 75, 80-82,
    99
    as greenhouse gas, 45-46
    Permian Extinction, contribution
    to, 37, 99-102
    photosynthesis, removal through,
    32-36, 45, 80, 106, 124
    predictions, 99, 101
    temperature, effect on, 46

climate change
    in fossil record, viii, x, 15, 106-107
    natural occurring, 116-117, 137-
    139, 156-157
    rapid, 15, 98, 110, 116
    regulatory factors, 18-23
    skepticism and debate of, 3-4, 41-
    42, 115-117

    solutions (see human)

clouds. See also water vapor
    albedo factor, 59
    as greenhouse factor, 44
    photochemical smog, 44

coal
    carbon content, range of, 119-122
    coal-fired electricity, 111, 139-
140,   149, 161-162
    formation, in Permian, 103

Cretaceous/Tertiary Extinction, 22, 87-90

dinosaurs, 22-23, 86-90

Earth
    climatic cycles, natural, 18-19,
    116, 137, 153-154
    extinction (see extinction)
    life (see life)

energy entrapment, 19-21, 28
    albedo, 55-59, 129
    carbon, 119

carbon dioxide, 45
    clouds, 44
    limiting, 129-130
    methane, 46-47
    nitrous oxide, 53
    oceans, 27-28, 44
    water vapor, 43-44

energy source, alternative, 68
    hydro, 67-68, 147-148, 159
    geothermal, 68, 148, 161
    nuclear, 159-161
    solar, 147-149
    wind, 147-149

extinction. See also life. See also Permian

extinction. See also Cretaceous/Tertiary

extinction
    dinosaurs, 22-23, 86-90
    species lost, 14-15, 84-90, 102-107
    triggers of, 36-37, 102-103, 111

fossil fuel. See carbon

fossil record, 12-13
 in Earth's early history, 82-84
 climate change triggers, evidence
 of, viii, x, 16, 107, 116
 of ocean conditions, 10, 36
 Permian fossil, 106-107

gas hydrate. See methane

geothermal
 as energy source, alternative, 68,
 148, 161

glaciation
 ice age, 81-85, 92
 Milankovitch cycles, affected by,
 64
 temperature, effect on, 23

"The Great Dying." See Permian

greenhouse gas, 42-55. See also carbon
 dioxide. See also methane. See
 also nitrogen oxide. See also water
 vapor
 in Earth's early history, 92

Hadley Circulation/Cells, 30, 99

human. See also Kyoto Protocol. See also
 Intergovernmental Panel on Cli
 mate Change (IPCC)
 carbon dioxide, contributing to,
 45
climate change
 adapting to, vii, 4-6, 15, 154-155
 contributing to, 116
 controversies, 116
 initiatives and policies, 55, 135,
 144-145
 solutions, 136-173
 energy source, alternative (see
 energy source, alternative)
 energy substitution, 139-142
 emission sequestration, 142-144
 greenhouse gas reduction, 144-
 147
 methane conversion, 157-158
 nitrous oxide, contributing to, 52-
 53

 population growth and activities,
 64-69, 136-137
 rainforests, effect on, 12-13

humidity. See water vapor

hydro
 as energy source, alternative, 67-
 68, 147-148, 159

ice age. See temperature. See glaciation

intelligent design, ix

Intergovernmental Panel on Climate
 Change (IPCC), 52, 99-101, 117

Kyoto Protocol/Treaty, 54, 144-145

life. See also extinction. See also human
 complex forms, x, 77-80
  in jeopardy, 104
 dinosaurs, 22-23, 86-90
 mammals, 89-90
 marine phytoplankton, 32-34
 marine carbonate shelled, 128-
 129, 155
 photosynthetic, 32-33, 78-81, 124-
 125, 155
 species, lost, 14-15, 84-90, 102,
 107

meteorite
    impact, effect of, 18, 22, 24
    life-formed carbon, evidence of, 77

methane/methane clathrate, 9, 46-49
    collapse, 90, 110-111, 154, 156-157, 168
    conversion, 157, 162-163
    as greenhouse gas, 46
    formation of, 125
    as natural gas, 46
    Permian levels, 107-111

Mikankovitch cycles, 62-64

nitrous oxide/nitrogen
    energy entrapment, 52-54
    molecular structure, 118
    photochemical smog, 44

nuclear
    as energy source, alternative, 159-161

ocean, 6-10. See also sea level
    albedo, effect on, 56
    carbon dioxide (see carbon dioxide. see carbon)
    carbonate shelled organisms, 128-129
    chemistry, 32-37
        during Permian, 102
    climate, effect on, 6-10, 28
    currents, 27-28
        Alaska, 32
        during Permian, 98, 110
        thermohaline circulation, 7-9
    energy entrapment, 27-28
    greenhouse gas, effect on, 9
    methane clathrate (see methane)
    oxygen, regulation of, 32
    photosynthetic organisms (see under life)

oxygen
    banded iron, formation of, 78, 81
    in Earth's early history, 78-81
    molecular structure, 118
    ozone (see ozone)
    Permian levels, 16, 105-106, 110
    photosynthesis, increase through, 32-34, 45, 78-81, 106

ozone. See also oxygen
    depletion of, carbon caused, 55
    as greenhouse gas, 50-51
    origins of, 79
    photochemical smog, 44

Paleocene/Eocene Thermal Maximum (PETM), 90

paleontological record. See fossil record

paleontology
    value of, x-xi, 17, 23

Permian extinction, 16, 23, 85-86, 98-111
    carbon dioxide levels, 99-102
    coal formation, 103
    methane levels, 107-111
    oxygen levels, 105-110
    species lost, 97-98, 102-107
    trigger of, 37, 102-111

photosynthesis
    carbon dioxide, conversion of, 32-34, 45, 120-123
    in Earth's early history, 32-34, 45, 78-81, 106
    phytoplankton, 33-34, 106

plants, 10-13. See also rainforests. See also photosynthesis
    agriculture, effect of, 53, 59
    in coal formation, 103, 119-123
    deforestation, effect of, 45
    in Earth's early history, 86, 89, 98, 103-104
    grass, 89, 122-123

rainforests, 10-13, 122
    albedo, effect on, 12, 59, 127-128
    carbon dioxide, elimination of, 127-128
    climate, effect on, 127-128
    status of, 10-12

sea level. See also ocean
    fluctuations, 4-7, 98-99
    during Permian, 98-99
    temperature, affected by, 5
    solar. See also energy entrapment
    as energy source, alternative, 147-149
    Milankovitch cycles, 62-64
    sunspot activity, 60-62

species. See under life

stromatolites, 78

temperature
    albedo, affected by, 41, 55-59
    atmospheric composition, affected by, 18-19, 41-55
    carbon dioxide levels, affected by, 46, 115
    current, status of, 92
    in Earth's early history, 75-77, 81-86, 89-93
    glaciation, affected by, 81-85, 90-91
    meteorite impact, affected by, 18
    oceans, interaction with, 27-28
    sea level, effect on, 6-9
    sunspot activity, affected by, 60-62
    volcanic event, affected by, 21-22
    water vapor, regulation of, 43-44

thermohaline circulation, 7-9. See also under ocean

water vapor, 30, 43-44
    in clouds, 43-44
    in Earth's early history, 77
    evaporation of, 30-31
    as greenhouse gas, 43-44
    Hadley circulation, 30
    precipitation, 31
    temperature, effect on, 77

wind
    as energy source, alternative, 147-149

volcanics
    climate, effect on, 45, 102, 137
    during Permian, 99, 102, 107
    Siberian Traps, 99, 102
    temperature, effect on, 21-22, 82

Illustration Credits